Encyclopaedia
of
MICROSCIENCE, TECHNOLOGY AND ENGINEERING

Encyclopaedia of
Microscience, Technology and Engineering

Aditi Mahalingam

ANMOL PUBLICATIONS PVT. LTD.
NEW DELHI-110 002 (INDIA)

ANMOL PUBLICATIONS PVT. LTD.

Regd. Office: 4360/4, Ansari Road, Daryaganj,
New Delhi-110 002 (India)
Ph.: 23278000, 23261597

Branch Office: No. 1015, Ist Main Road, BSK IIIrd Stage
IIIrd Phase, IIIrd Block,
Bangalore-560 085 (India)
Tel.: 080-41723429
Visit us at: www.anmolpublications.com

Encyclopaedia of Microscience, Technology and Engineering

First Edition, 2009

ISBN 978-81-261-4114-2

PRINTED IN INDIA

Printed at Balaji Offset, Delhi.

Contents

Preface

Microtechnology is the popular term for the construction and utilisation of functional structures with at least one characteristic dimension measured in Nanometers. Such materials and systems can be rationally designed to exhibit novel and significantly improved physical, chemical, and biological properties, phenomena, and processes because of their size. Currently known nanostructures include such remarkable entities as Carbon Nanotubes, proteins, DNA, and single-electron transistors that operate at room temperature. Rational fabrication and integration of Microscale materials and devices herald a revolutionary age for science and technology, provided we can discover and fully utilise their underlying principles.

Micromaterials encompasses a wide range of areas and possibilities, and these are not just in the future. They are being incorporated into products and services today but the scope to do more is enormous. The more you examine the possibilities that nanotechnology opens up, the more fascinating the implications become. So it is no wonder that the attention of business, manufacturing, investors, scientists and politicians is focused on the progress being made in this growing area of science.

This encyclopaedia is a fascinating overview of nanotechnology which provides broad coverage of the basics of Microtechnology and its applications, with an eye toward giving students and researchers in different areas an appreciation of Microtechnological developments outside their own fields of expertise. Rather than focusing on the latest developments in nanotechnology, it use examples of research in many fields to focus on the diversity of its applications.

Chapter 1

Microtechnologies

Microtechnologies are a part of nearly every aspect of modern life. The border between the two technologies is fluid; microelectronics has been conducted on a scale measured in nanometers for some time. Applications and inovative products based on micro- and nanotechnologies are not limited to electronic components: Products on the market range from dirt shedding paints and scratch-free surface coatings to mechatronic systems such as the accelerometers which trigger airbags.

Clinical applications such as nano particles are used in cancer treatment. Components such as biochips and micro fluid systems are essential to modern pharmaceutical research. Not only continuous miniaturization alone, the integration of mechanical, electronic and optical components in the smallest possible space is an important trend in all fields of endeavor.

The advances in miniaturization come at a tremendous price. In order to manage the enormous technological and financial commitment, research is ongoing in all branches of micro- and nanotechnology to replace the current top-down method, which was adapted from solid state technology. This older technology employs lithography on the macroscopic level to structure processes.

The "bottom-up" procedure which holds promise as a replacement makes use of principles such as self-organization and molecular recognition (similar to biological systems and already in use in supramolecular chemistry) to create nanostructures. Both the latest discoveries and methods arising from the field of nanophysics are being applied successfully in, for instance, medicine and the life sciences. Additionally, concepts and tools from the fileds of biology and chemistry are leading to further minaturisazing and systems integration.

Nanotechnology is the interface technology of the 21st century.

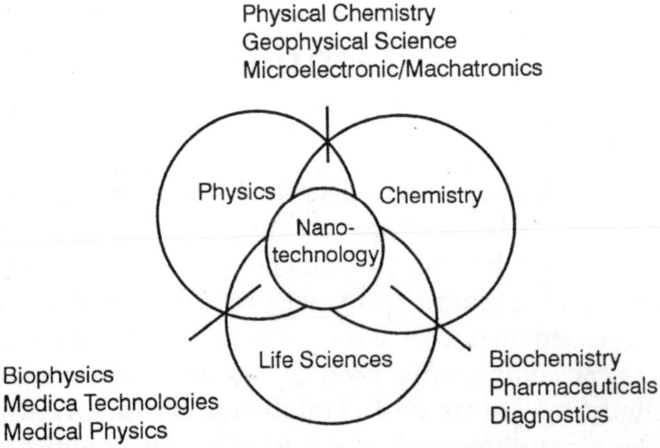

The objective of the Master's Studies in Micro- and Nanotechnologies at the Munich University of Applied Sciences is based on this interface between the natural and engineering sciences and to promote knowledge of the products, research and applications for these smallest of structures and systems. (The AFM has become an essential tool in the research of the smallest structures).

Fig. AFM (Atomic Force Microscope)

Traditionally, the semi-conductor industry has been one

of the most important sectors for graduates in the fields of microtechnology Unfortunately, the semi-conductor sector has been dependent upon economic cycles.

This may be less of a problem with nanotechnology in the future.Today, microtechnology and, especially, nanotechnology have found application in nearly every branch of industry.

There is an increasingly broad range of opportunities for graduates of our course of study in Micro- and Nanotechnoly in both private industry and goverment sponsored research. For the field of nanotechnology alone, there is an estimated global market share of one billion US dollars already by 2015 and a stude for the BMBF sees an annual need for professionals competent in the field of nanotechnology in Germany alone at between 5,000 and 8,000.

Gradutes of the course of study Micro- and Nanotechnlogy at the Munich University of Applied Sciences are to be found in all the most important sectors of micro- and nanotechnology.

These include the fields of semi-conductors and electronics, in the IT sector, aerospace and space flight, the automobile industry, the chemical industries, biotechnology, pharmaceuticals and heath industries, optical and laser based technologies as well as at universtities and government research institutions.

The professional responsibilities of these graduates cover the entire range of careers and responsibilities in these high-tech fields of endeavor.

The career opportunities include independent planning, coordination and supervision of research and development projects, management functions and industrial consulting. The interdisciplinary nature of this course of study provides the graduates with the best possible qualifications for the increasingly important interfacing positions of modern industry.

Approximately one-third of the graduates go on from their successful completion of this course of study to doctoral and post graduate work at other universities.

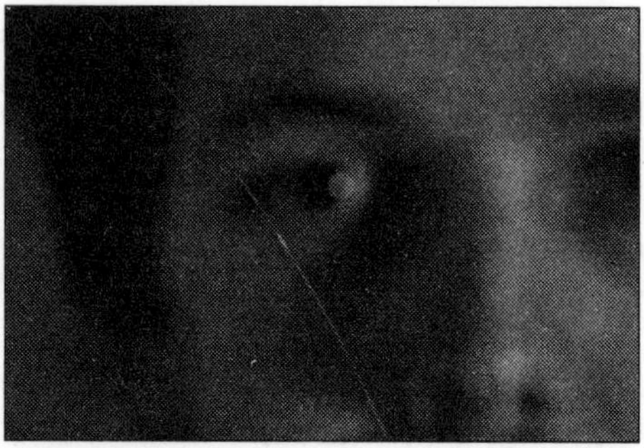

Fig. Application of Microtechnology

MICRO- AND NANOTECHNOLOGY IN MUNICH

Both academically as well as industrially, the metropolitan area of Munich has become established as one of the most important locations for the fields of micro- and nanotechnlogy in Germany and Europe.

Academically, advanced research facilities such as the Centre for NanoScience, the "Excellence Network Nanobiotechnology", the NanoTUM and the Nanosystems Initiative Munich, which is supported under the Excellence Initiative of both the Federal and State governments and of which the Department of Precision- and Micro-Engineering, Engineering Physics at the Munich University of Applied Technologies is also a member.

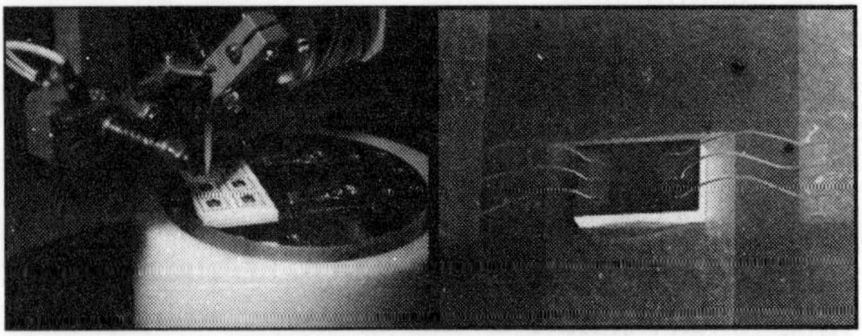

Fig. Instruments of Microtechnology

Outstanding industrial firms such as Siemens, Infineon Technologies, Qimonda, EADS, Olympus as well as a number of small and mid-sized firms are located in the metropolitan area. Over the last years, a significant number of successful start-up firms have arrisen from the academic environment Munich provides.

The Department of Precision- and Micro-Engineering, Engineering Physics, together with our instructors, has outstanding connections to the industry providing graduates of the Course of Study Micro- and Nanotechnolgy with excellent networking opportuntites.

SAMLLER IS BETTER

A dime-sized amplifier makes fibre-optic communications faster and clearer. A portable DNA analyzer helps detect and identify organisms in the field, including human remains and biological warfare agents. A tiny gripper inserted in a blood vessel treats aneurysms in the brain to ward off potential strokes. What do these technologies have in common? Each one is smaller than any comparable product, opening up a host of new applications. And each originated in Lawrence Livermore National Laboratory's Microtechnology Centre.

In the late 1960s, Livermore scientists and engineers began making miniature devices for high-speed diagnostic equipment required for nuclear tests. For many years, before the development of Silicon Valley and the ready availability of microchips for a broad array of uses, Laboratory engineers fabricated chips to their own specifications for high-speed switches, high-speed integrated circuits, and radiation detectors. By the early 1980s, Livermore was fabricating thin-film membranes for use as x-ray windows in low-energy x-ray experiments, as x-ray filters, as debris shields for the Extreme Ultraviolet Lithography Programme, and as targets for high-energy electron experiments in which x rays are generated.

These passive microstructures have been applied to dozens of projects. They have served as diagnostic devices for Livermore's Nova laser experiments and will do the same for experiments at the National Ignition Facility (NIF). Another microstructure, a novel thin-film window developed by Glenn Meyer and Dino Ciarlo, plays a critical role in a new, more efficient electron-beam system for processing inks, adhesives, and coatings. Laboratory scientists, led by Booth Myers and Hao-Lin Chen, teamed with American International Technologies Inc. of Torrance, California, on this project and won an R&D 100 Award in 1995. Conventional electron-beam processing systems are inefficient, delivering about 5 per cent of the beam's energy to the polymer being cured. With this new window, efficiencies greater than 75 per cent were achieved. The team also recently won a 1997 Federal Laboratory Consortium Award for Excellence in Technology Transfer. In the mid-1980s, Livermore began combining micro-optical devices with microelectronics for extremely high-speed, fibre-optic data transmission.

Photonic devices have since found their way into many microtechnologies that incorporate optical fibers for transmission of laser light. Livermore stopped fabricating silicon-based electronic circuits when commercial microchips became available in almost every configuration imaginable. But invention by no means stopped.

Today, the Microtechnology Centre, now headed by physical chemist and engineer Ray Mariella, invents and applies microfabricated components, including photonic devices, micro-structures, and microinstruments, to directly support Laboratory projects in science-based stockpile stewardship, nonproliferation, and biomedical research. At any given moment, the Centre has about 25 projects in the works. The Center's major recent and ongoing projects are highlighted here.

The Center's state-of-the-art fabrication facilities are centered in a building whose location was selected because the area had the smallest vibrations within the Laboratory site, permitting the high-resolution microlithography that the Centre performs. Microdevices can be fabricated there in any of three material systems:

- Silicon and silicon compounds for microstructures and microelectro-mechanical systems applications.
- Gallium-arsenide for photonics applications.
- Lithium niobate for electro-optic applications, such as phase and amplitude modulators.

The Centre has the equipment and infrastructure needed for lithography, etching, diffusion, wafer bonding, and thin-film deposition and vacuum techniques. Its dry laboratories are used for surface inspections, packaging, and electrical and optical device testing.

Groundbreaking recently took place for an addition that will increase clean-room space by 65 per cent. The backbone of the Centre is an interdisciplinary group of about 50 electronics, mechanical, chemical, and biomedical engineers, physicists, and technical support personnel. According to Mariella, "Ideas, technologies, and capabilities are shared at frequent brainstorming sessions, so staff can find solutions to programmatic problems quickly."

PUTTING LIGHT TO WORK

Photonics work at Livermore got its start from the need to obtain remote, highly accurate measurements at nuclear weapons tests. Photonic systems—which manipulate and exploit light for control, communication, sensing, and information display—were the ideal solution because signals can travel on them for long distances at the speed of light with very little power loss.

After several years of development, Livermore successfully fielded its first photonic system for measurement of ionizing radiation from a nuclear weapon at the Nevada Test Site in 1989. This system made available very high-resolution data that conventional measuring techniques could not deliver. In 1991, Livermore was awarded the DOE Weapons Excellence Award for this work.

A photonic network is typically made up of optical fibers, waveguides, amplifiers, receivers, wavelength selection elements, and modulators, sometimes all on a single chip. One of the Laboratory's contributions to the photonics field has been its novel application of silicon micromachining capabilities, which have been critical to packaging photonic components in a cost-effective manner.

In 1994, Mike Pocha, Dan Nelson, and Ted Strand won an R&D 100 Award for the development of a silicon "microbench" that reduces the time needed to align and connect the optical fibre in photonic components.

Because of the submicrometer alignment tolerances, the standard manual process was extremely time consuming and therefore expensive. The team's technique provides just enough heat to melt the microdrops of solder needed to make the connection, allowing a rapid manual alignment and connection of the fibre to a laser diode or a lithium niobate modulator in less than five minutes and reducing the cost for this work by 90 per cent.

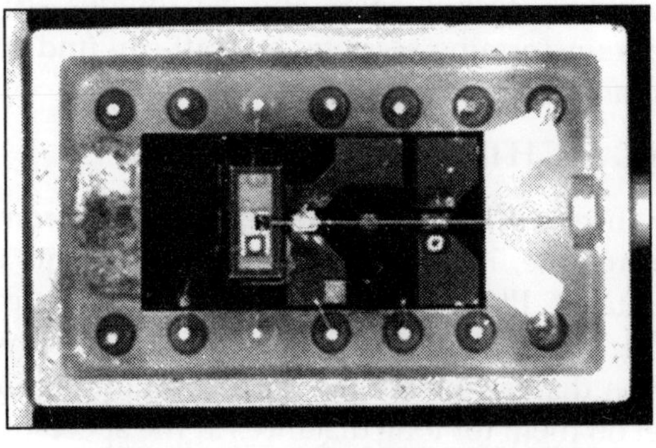

Fig. With the Silicon Microbench

Two polysilicon heating elements and gold solder attachment bases provide the means for attaching the optical fibre, while the fibre is held in position, current is passed through the heater to reflow the solder, which wicks around the metalized fibre without disturbing the alignment. This new method avoids thermal shifts and simplifies the alignment process.

Photonic devices are finding their way into two different parts of the Laboratory's science-based stockpile stewardship programme. One is ultrascale computing, which soon will be used for simulation of nuclear tests; the other is diagnostics for NIF.

Computing on a large scale requires the cooperative action of thousands of microprocessors sharing tremendous volumes of data. This data sharing demands a communication "fabric" of very high bandwidth and low latency (short time delay) to enable the microprocessors to function without waiting for data. Optical fibers are an ideal medium provided that the optical signals are sufficiently amplified so that there are enough photons to go around for the many receiver nodes. Existing amplifiers had problems: erbium-doped fibre amplifiers are bulky and expensive, and conventional semiconductor optical amplifiers produce too much crosstalk at transmission rates above 1 gigabit per second.

To solve this problem, Sol DiJaili, Frank Patterson, Jeff Walker, Robert Deri, William Goward, and Holly Peterson developed a miniature, low-cost semiconductor optical amplifier. They applied state-of-the-art microfabrication techniques including molecular beam epitaxy to grow device wafers and chemically assisted ion-beam etching to make the device structures. The team won a 1996 R&D 100 Award for their new device, which can be used not only for computer interconnections but also in wide-area networks and for transmitting visual images.

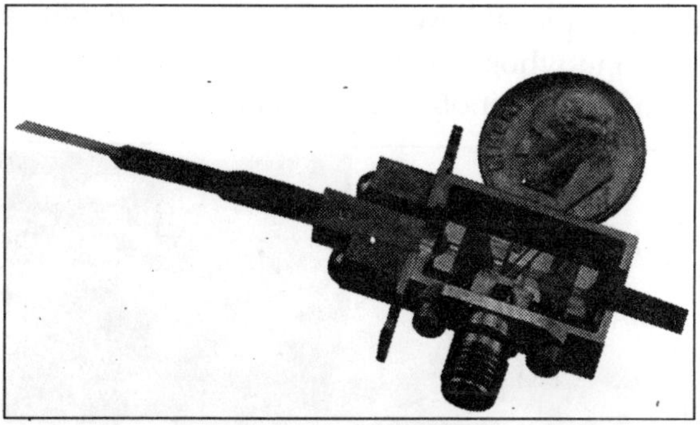

Fig. This Dime-Sized Semiconductor Optical Amplifier Reduces Crosstalk and Noise in Fibre-Optic Communications.

They then replaced the standard gain medium inside the waveguide of the amplifier with a tiny vertical-cavity surface-emitting laser and took advantage of some basic laser properties to reduce crosstalk by a factor of 10,000.

The photons' stimulated emission in the gain medium when lasing occurs acts as a clamp on the signal gain, eliminating the fluctuation. Signal channels at multiple optical wavelengths can pass through the waveguide with virtually no crosstalk among these channels. The lasing action also speeds recovery time of signals through the waveguide, from a billionth of a second to about 20 trillionths of a second. Thus, the amplifier can successfully track the amplification of a serial bit stream at very high bit rates.

The Microtechnology Centre is also applying photonics technology at NIF. Because NIF's 192 laser beams will be aimed at such small targets (about the size of mustard seeds), NIF will need much faster radiation diagnostics than those used at the Nevada Test Site.

Mike Pocha and Howard Lee are developing photonic radiation sensors that will modulate an optical beam in response to an ionizing radiation input and then record it using single-shot optical samplers having a response time of 100 femtoseconds (quadrillionths of a second). Pocha and Lee are investigating the use of waveguides made of nonlinear optical material to perform the extremely high-speed signal gating required to sample at these high data rates. (A nonlinear optical material is one whose index of refraction can be changed by the introduction of another light beam.)

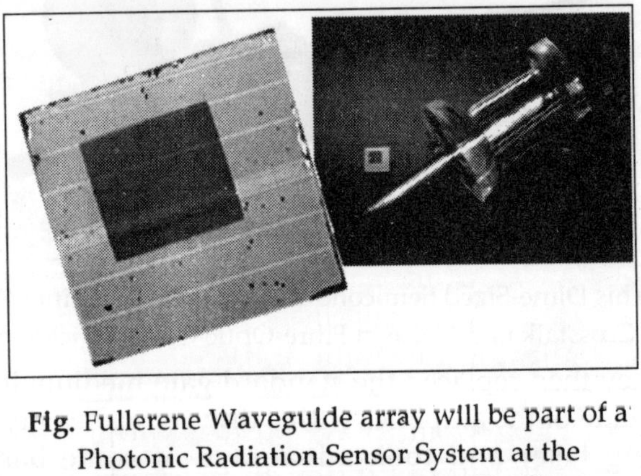

Fig. Fullerene Waveguide array will be part of a Photonic Radiation Sensor System at the Laboratory's National Ignition Facility

These nonlinear gates will be capable of switching sequential slices of the radiation-modulated optical signal into an array of relatively slow-speed optical detectors. A material that may have the right nonlinear properties is fullerene (C_{70}), whose discoverers recently won the Nobel Prize in Chemistry. (Fullerene is a van der Waals crystal with molecules shaped like Buckminster Fuller's geodesic domes, hence its name.) The world's first fullerene waveguide array, which is still undergoing development.

Work continues on this project so that a fully functional system will be on line in time for the testing of NIF, which is scheduled for 2002.

Analyses in Miniature

Analyzing DNA, testing for HIV, and identifying pathogens and poisons used in biological and chemical warfare all require sampling a range of products. Supporting the Laboratory's bioscience research programme and its nonproliferation efforts, the Microtechnology Centre has developed several cutting-edge microdevices that facilitate biological and chemical sampling and analysis in the field, allowing real-time detection. Because some samples cannot survive transport from the field to a remote laboratory, field analysis is often the only solution.

Biological and chemical sampling with micro-instruments offers other advantages as well. Smaller instruments have lower power requirements. Highly integrated and automated sample handling systems usually result in improved productivity and less sample contamination. Also, because analytical diagnostic procedures sometimes produce hazardous wastes, smaller systems mean less waste. However, extremely small-volume chambers require greater sensitivity in order to identify the extremely small trace samples.

A team led by Ray Mariella has patented a new system that eases alignment and increases the accuracy of flow cytometry. Flow cytometry is a powerful diagnostic tool used to characterize and categorize biological cells and/or their contents, such as DNA.

It is used by laboratories throughout the world for blood typing and for testing for a wide variety of diseases and viruses, including HIV. The cells flow in single file in solution while the experimenter directs one or more beams of laser light at them and observes the scattered light, which is caused by variations in the cells or DNA.

Instead of using a microscope lens or an externally positioned optical fibre as a detector, Mariella's system uses the flow stream itself as a waveguide for the laser light, capturing the light and transmitting it to an optical detector. Alignment simply requires lining up the light source onto the flow stream and placing the detector into the same stream. With this system, measurements are up to three times as accurate as those taken with conventional systems.

At international joint field trials last fall at Dugway, Utah, the new flow cytometer performed extremely well detecting simulated biological warfare agents. Participants from the U.S., the U.K., Canada, and France used a variety of instruments to detect four simulants. The Livermore flow cytometer detected 87 per cent of all the unknowns with a false positive rate of just 0.4 per cent.

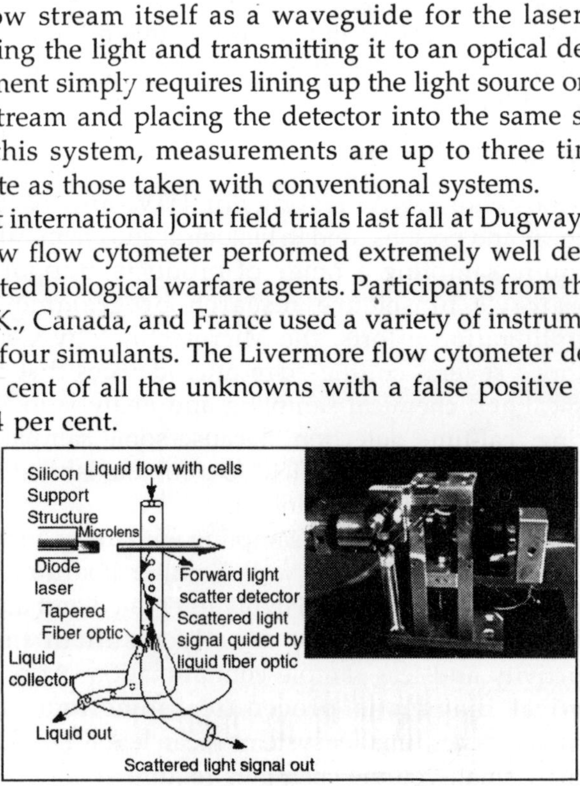

Fig. (a) Schematic and (b) Demonstration Model of the New Flow Cytometer

As a cell passes through the laser beam on the left, the cell simultaneously scatters (reflects) the laser light into the forward-light-scatter and the right angle-scatter detectors. The scatter pattern reveals the cell's size and internal structure.

Dino Ciarlo, Jim Folta, and William Benett developed a miniature sample injector that can be used in Mariella's flow cytometer. Flow channels are formed by etching three silicon wafers and bonding them into a single chip. Fluidic connections are made to this injector chip via a plastic block. A thin gasket, laser-cut from an elastomeric material, forms a seal between the silicon chip and the block. The front edge of the silicon chip remains exposed to allow fluid to freely exit through an edge port into the optical detection system.

A Laboratory team headed by M. Allen Northrup has developed an portable DNA analyzer that is small enough to fit in a briefcase. It is also fast. This unit, the world's only battery-powered DNA analyzer, moves analysis out of the laboratory for the first time. Folta and Benett developed a disposable polypropylene liner for the analyzer's tiny heated chamber where the polymerase chain reaction occurs. The liner facilitates rapid reuse of the chamber and eliminates tedious cleaning and possible contamination. The entire chamber is a tiny chip of silicon. As the reaction progresses, the team uses a fluorescent signal to analyse the DNA to determine whether it matches that of a particular subject.

Fig. DNA Analyzer and Computer System Fit in a Briefcase

LLNL has delivered one of these units to the Department of Defence Armed Forces Institute of Pathology, which is using it to quickly identify human remains in the field, test food and water for contamination in remote locations, and identify pathogenic bacteria on the battlefield.

The Microtechnology Centre is also supporting the

Laboratory's DNA sequencing work for the Human Genome Project. As described in the research highlight High-Speed DNA Sequencing, the Centre has developed etched and bonded microchannel glass plates to speed up the sequencing process. A patent is pending on the new bonding process. Conrad Yu of the Centre is participating in work on a miniature, portable, low-power gas chromatograph to support the Laboratory's programme in nonproliferation to counter the spread of chemical weapons.

Gas chromatography is a proven method for identifying liquid or gas species, with detection sensitivities as high as parts per billion. Conventional gas chromatographs, however, are several cubic feet in size and typically take about 20 minutes to analyse a gas sample.

A mini unit works faster, often requiring just one minute to complete an analysis, and would be very useful to carry into an area where chemical weapons or other poisonous gases are suspected to have been used. Someday this unit could also be used at home for sniffing out radon gas. Yu has developed a micromachined, silicon sample injector about the size of a little fingernail.

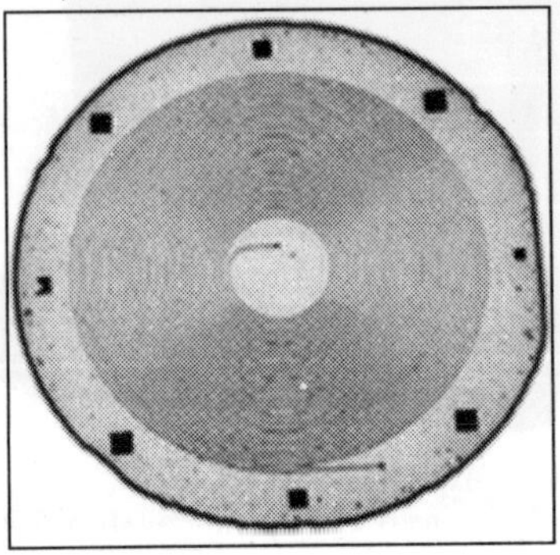

Fig. The Column for the Miniature gas Chromatograph has been Reduced to two Silicon Wafers Bonded Together

He has also reduced the size of the chromatograph's column where the various elements in the sample are separated before being directed to the detector where they are identified. The column has been reduced from 1.6 liters (100 cubic inches) for a laboratory-sized unit to a coil etched on a silicon wafer. A circular column 100 micrometers wide and several meters long is etched on two silicon wafers that are bonded together. The entire instrument occupies about 0.16 liter (10 cubic inches).

Microtools for Better Health

A new surgical tool for treating aneurysms, the silicon microgripper is about the size of two salt grains—1 by 0.2 by 0.4 millimeters. With guidance from researchers at the University of California, San Francisco, Abraham Lee, M. Allen Northrup, and Peter Krulevitch developed this micro-electromechanical device.

The microgripper is like a tiny hand that surgeons can use to place clot-inducing agents to fill an aneurysm. A surgeon may also use it to perform minimally invasive in vivo biopsy or catheter-based endovascular therapy. Nonmedical uses include assembling small parts for manufacturing and remote handling of small particles in extreme environments, such as high or low pressures and hazardous fluids.

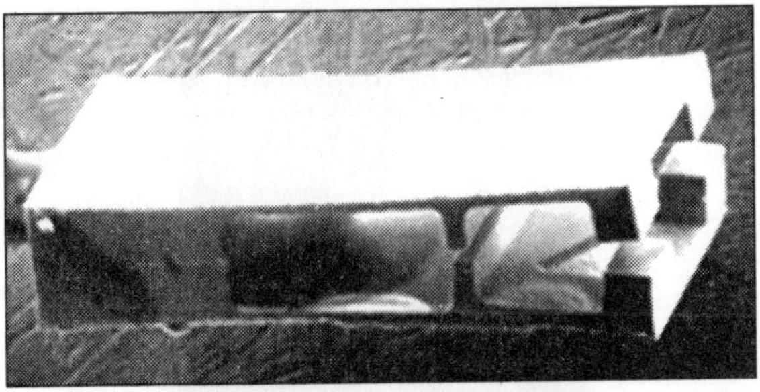

Fig. Microgripper, the size of two Grains of Salt, is the First in a Series of New Surgical Microtools for Healthcare Technologies

A key to the microgripper's effectiveness is a thin-film microactuator that is fabricated from a shape-memory alloy (SMA). At low temperatures, SMAs are easily deformed, but when heated, they recover their original shape.

This reversible transformation forms the basis for shape-memory actuators, in which a biasing force, produced by a spring, for example, deforms the SMA element at low temperature, and the SMA element overcomes the bias when heated. For the microgripper, the team developed a sputter-deposited shape-memory actuator of nickel-titanium- copper, with a transformation temperature just above body temperature.

The microgripper is inserted into a blood vessel in the closed (deformed) position. Through a thin wire connected to the microgripper, an electrical current of 0.1 milliamps activates the actuator, deflecting each arm up to 55 micrometers and returning the gripper to its undeformed (open) position.

As it cools, the gripper will open again. (A patent was recently issued for another microgripper made of plastic with a balloon actuating system.)

Lee, Julie Hamilton, and Jimmy Trevino have also built a low-leakage, high-efficiency microvalve. Effective microvalves are an important link in creating miniature total analysis systems that can be used for drug delivery and bioanalytical instrumentation.

Fig. The 200 micrometer-long Microvalve is shown in the open Position, Revealing an Etched Hole

Nonmedical applications for the microvalve include fluid injection analysis, chemical processing and analysis, and atmospheric and temperature control equipment. In this design, an electrode is sandwiched between two polyimide films with different coefficients of thermal expansion.

(Polyimide is a flexible plastic material.) Delivery of less than 1 milliwatt of power causes the "cantilever" to clamp down, sealing an etched hole beneath it.

MICRO BTS

Micro BTS (Buddy Tracking System), a location based product, is geared towards delivering highly personalized services to the present generation of mobile telephony users who are constantly on the move. Micro BTS offers a unique service that enables cell phone subscribers to share their real-time location status and messages, with their friends from a mobile phone.

Using Global Positioning System (GPS) technology, the application automatically updates the location of a user's "buddies" and displays the information directly on a map on the phone.

Alerts can also be sent to notify subscribers when a friend is near. This product offers a mix of social networking with the help of long messaging system and location-based services. So the ability to automate becomes a really valuable service.

Micro BTS enables the user to stay in touch when they travel to another region or country. It is like a community-based approach where every user is connected. Using this system, one can share one's position with his/her friends and family, creates and joins groups, records and reviews his/her track history.

From the corporate point of view Micro BTS can be used as an aid to increase the efficiency of the organizations with proper utilization and management of its sales, marketing and service staff, because success of an organization depends on the competencies of the sales staff and proficiently managed team definitely adds on value in terms of both efficiency and revenue to the organization.

Micro BTS can also be used in a very proficient way with law enforcement agencies, in tracking the movement of any suspicious persons and utilizing the law enforcement officers to their best of ability minimizing the time involved in communication.

MICROREACTOR

The microreactor concept was realised by Ponton in 1993, describing a hypothetical miniaturised disposable chemical plant where a parallel scale out approach could produce significant batch quantities of product whilst allowing safe 'at point of use' manufacture of HCN for example.

Recently, Chambers reported the selective fluorination of various organic species, however the system described used a nickel microreactor requiring reagent delivery via syringe or syringe pumps. In this study solvent and sample pumping is achieved by electroosmotic flow (EOF) rather than hydrostatic displacement as used by Chambers. EOF has been selected for this particular application as it offers a simpler sample handling methodology with greater control and selectivity of the reaction chemistry.

The work described here is based on the optimisation of a Wittig synthesis which is used in industry for a variety of syntheses e.g. vitamin A producing a C-C link with the formation of olefinic double bond.

The reaction was selected for two reasons, firstly to demonstrate synthetic application of microreactors and secondly, to establish the potential to perform combinatorial chemistry in such devices.

Methodology has illustrated the potential of microtechnology to perform catalysed organic reactions. The Wittig synthesis however offers simple solution chemistry, whilst offering the advantage of producing a coloured intermediate (ylide), which allows visual solution profiling and diffusive mixing to be achieved.

The flow injection microreactor was fabricated using photolithographic techniques. Microporous silica frits (MPSF) were generated using the method outlined and positioned in microchannels as indicated in figure. The channels were then sealed by annealing to the patterned base plate to a 17 mm thick top plate (680 °C) in a microwave furnace. The top plate included 3 mm internal diameter pre-drilled holes aligned at the ends of each channel to act as reservoirs and electrode supports.

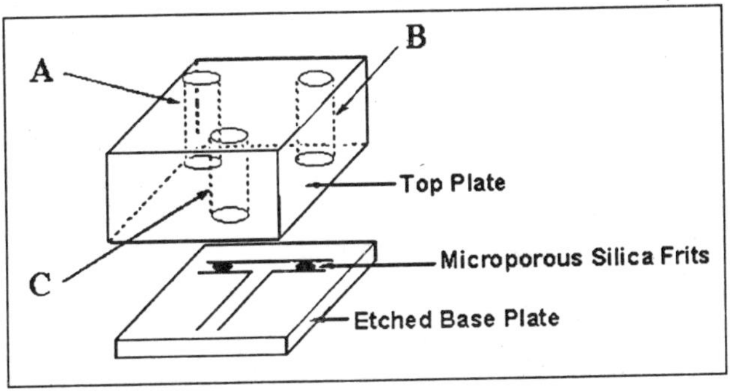

Fig. Schematic Diagram of the 'T'-shaped Manifold used in the Reactor for the Wittig Synthesis

The synthetic method used was adapted from that previously reported by Hughes et al. The continuous flow micro-reactor was operated in the following manor. Prior to synthesis, the microchannels were primed with dry, degassed methanol (MeOH) removing any excess moisture from the channels and reducing the side reaction of hydrolysis.

A standard solution of 2-nitrobenzyltriphenylphosphine bromide (80 microlitres, 0.01 M) in MeOH was added into reservoir A of the microreactor. Methyl-4-formylbenzoate (0.02 M) was basified with sodium methoxide (0.3 ml, 0.5M standard solution concentration) and 80 microlitres of the premixed solution was introduced into reservoir B. MeOH (40 microlitres, dry, degassed) was introduced into reservoir C. Electrodes were then placed in each reservoir (A + B positive, C ground), which were then covered using laboratory film. Two methods of mobilisation were employed.

Mobilisation method 1: an external applied voltage (in the range of 100 – 700 V) was applied to channels A (+) and B (+) relative to reservoir C (-). This induced the continuous flow of methyl-4-formylbenzoate from B (+) to C (-) and 2-nitrobenzytriphenylphosphine bromide from A (+) to C (-). This method was used for the conventional batch methodology i.e. the aldehyde in excess for both synthetic and the combinatorial approaches.

Mobilisation method 2: involved the application of the external voltage (400V) to reservoir B (+) relative to C (-), inducing the flow of methyl-4-formylbenzoate. The periodic application of 400V between A (+) to C (-) efficiently injected reagent slugs of 2-nitrotriphenylphosphine bromide into the continuous stream of methyl-4-formylbenzoate. Mobilisation method 2 was used for the synthetic optimisation of 1:1 stoichiometry in the microreactor. The microreactor was run for 20 minutes, after which the reaction was quenched using dilute HCl before analysis. For both of the above methods, the total volume of the solutions in reservoir A and C at the end of each reaction was recorded and samples were taken for HPLC-MS analysis. The LC peak area was used to determine the solution yield of the reaction.

SYNTHETIC REACTION DEVELOPMENT

Using mobilisation method 1 the optimum reaction yield relative to the applied voltage or flow was obtained. At high voltage it was apparent that the reaction was not occurring in the reaction manifold (the purple ylide intermediate was observed in reservoir C), which resulted in no increase in yield compared with traditional batch reactions. At lower voltages (ie. less than 400 V), the purple intermediate was observed in the mixing channel of the reactor and the reaction yield was increased by approximately 40 per cent over the batch method, however the yield obtained was non-reproducible and yields of over 100 per cent were obtained (mainly at 100V). The reaction was optimised further at 400V.

Figure summarises the results obtained for the synthetic and combinatorial approach carried out at 400V for both the continuous (Method 1) and injection (Method 2) profiling. The graph summarises the results for methyl-4-formylbenzoate. The traditional batch yield using the basified aldehyde was 60 per cent (2:1 stoichiometry), however the microreactor using continuous flow of both reagents was 70 per cent (2:1), demonstrating a 10 per cent increase in the reaction yield. For stoichiometry of 1:1, the batch reaction yield was 48 per cent and the microreactor exhibited an 11 per cent increase.

Whilst the yield increase for the microreactor may not seem substantial, the time to perform the experiments was significantly reduced (20 minutes batch, <5 minutes microreactor). This aspect of microreactor technology is very powerful in synthetic chemical application where one may wish to see if a simple reaction is working. The general applicability of the microreactor was tested using four other aldehydes exhibiting different diversities.

As can be seen from the remainder of the graph, the 3-benzyloxybenzaldehyde and 2-naphthaldehyde exhibited very similar yields when compared with that of the batch synthesis however, the yield obtained for 5-nitrothiophene-2-carboxaldehyde (1:1) was 9 per cent lower than for the traditional scale. It must be noted that the four aldehydes tested used the optimised method for the methyl-4-formylbenzoate. This approach allows for the development of a combinatorial device allowing screening for a variety of compounds.

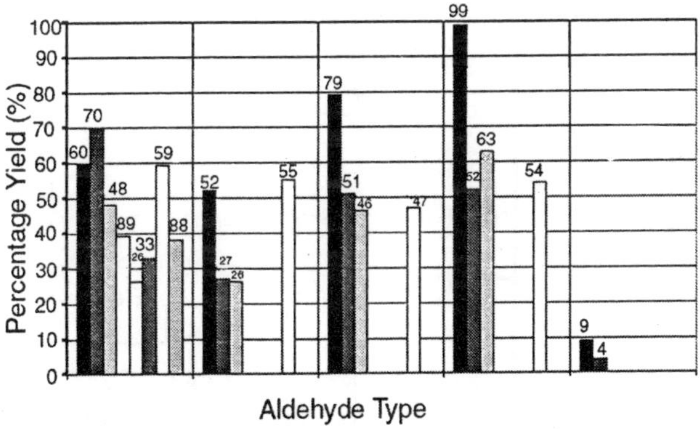

Fig. Result Summary for Combinatorial and Synthetic Experiments

COMBINATORIAL CHEMISTRY PROTOTYPE

Combinatorial chemistry allows the synthesis of a large number of organic species in which a large number of chemical combinations are achieved. The methodology enables the fast generation and screening of combinatorial libraries against biologically active target compounds ideal for the drug discovery process. Combinatorial libraries exhibit greater SAR (structure activity) and therefore increased likelihood of drug discovery success due to the immense libraries that can be synthesised using very few synthetic steps (fifty individual reactions coupled with fifty containers can synthesise a diverse library of 100,000 members in one week).

The speed of combinatorial analysis is already at a significant scale however, the attributes of microtechnology offer a highly selective method potentially increasing the number of potential members that could be synthesised per week. The following information describes the development of the optimised synthetic reaction outlined previously enabling the production of a semi-automated prototype combinatorial screening device.

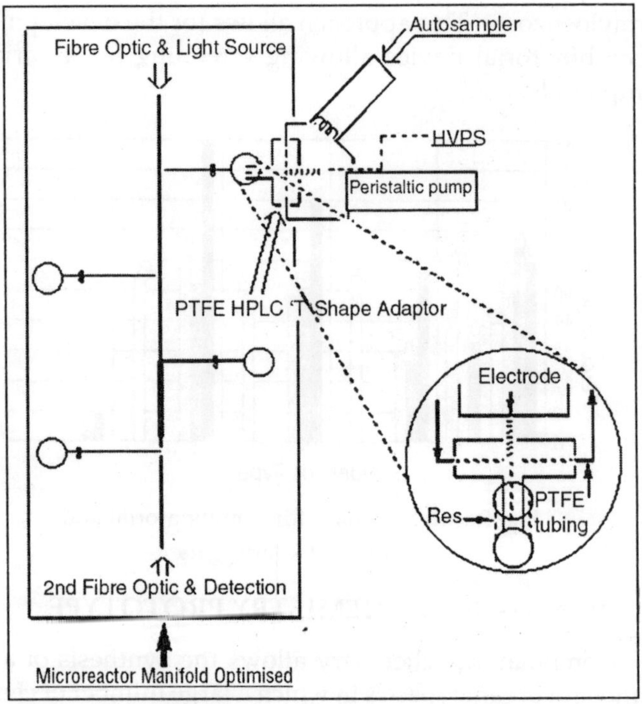

Fig. Schematic of Semi-automated Aldehyde Screening Device

Figure is a schematic of a simple semi-automated screening device. In this case the microreactor manifold is different to that earlier described allowing the positioning of the fibre optics leading from the halogen light source to the diode array detector. The reservoir (magnified) shows the positioning of a PTFE block that has been machined to fit the reservoir diameter. The block consists of an inlet and outlet section to which PTFE tubing is connected allowing the delivery/removal of solution/reagents.

The correct amount of solution/reagent can be administered into the reservoir by using a flow injection loop that allows the injection of the following profile, aldehyde, methanol, aldehyde, and methanol preventing any cross contamination. The reagents are delivered from each reservoir and react infront of the fibre optic detection system and then flow to waste.

New starting materials are introduced from the autosampler located in the flow injection manifold. The reaction and detection can be determined all on-chip however, for the system to be fully optimised, a secondary micro-total analytical system is required allowing product isolation and reagent recycling. The system described in Figure is currently under further investigation.

Microtechnology offers the potential for novel tools with which to control chemistry and enable unique insight into reaction chemistry to be achieved. For many synthetic chemical applications such devices will offer, for example, increased reaction yields and reduction in stoichiometry. Further the development of a semi-automated combinatorial screening device potentially offers a rapid screening device producing a vast and varied combinatorial library.

Chapter 2

Micro Molecular World

The sizes, forces, and general nature of objects in the molecular world. Building on this, we can get a better picture of where developments seem to be leading, a better picture of molecular manufacturing itself. To show the sizes, forces, and general nature of things in molecular manufacturing, we first invite the reader (and the reader's inquisitive alter ego) to take a second and final tour before returning to the world of present-day research. As before, the pre-1990 history is accurate, and the science isn't fiction.

THE SILICON VALLEY FAIRE

The tour of the molecular world showed some products of molecular manufacturing, but didn't show how they were made. The technologies we remember from the old days have mostly been replaced—but how did this happen?

The Silicon Valley Faire is advertised as "An authentic theme park capturing life, work, and play in the early Breakthrough years." Since "work" must include manufacturing, it seems worth a visit. A broad dome caps the park —"To fully capture the authentic sights, sounds, and smells of the era," the tourguide politely says. Inside, the clothes and hairstyles, the newspaper headlines, the bumper-to-bumper traffic, all look much as they did before our long nap. A light haze obscures the buildings on the far side of the dome, our eyes burn slightly, and the air smells truly authentic.

Pocket Libraries

The Nanofabricators, Inc., plant offers the main display

of early nanotechnology. As we near the building, the tour guide mentions that this is indeed the original manufacturing plant, given landmark status over twenty years ago, then made the centerpiece of the Silicon Valley Faire ten years later, when. With a few taps, we reset the pocket tour guide to speak up less often.

As people file into the Nanofabricator plant, there's a moment of hushed quiet, a sense of walking into history. Nanofabricators: home of the SuperChip, the first mass-market product of nanotechnology. It was the huge memory capacity of SuperChips that made possible the first Pocket Library.

This section of the plant now houses a series of displays, including working replicas of early products. Picking up a Pocket Library, we find that it's not only the size of a wallet, but about the same weight. Yet it has enough memory to record every volume in the Library of Congress–something like a million times the capacity of a personal computer from 1990. It opens with a flip, the two-panel screen lights up, and a world of written knowledge is at our fingertips. Impressive.

Nanofabrication

The next room displays more technology. Here in the workroom where SuperChips were first made, early nanotech manufacturing is spread out on display. The whole setup is surprisingly quiet and ordinary. Back in the 1980s and 1990s, chip plants had carefully controlled clean rooms with gowns and masks on workers and visitors, special workstations, and carefully crafted air flows to keep dust away from products. This room has none of that. It's even a little grubby.

In the middle of a big square table are a half-dozen steel tanks, about the size and shape of old-fashioned milk cans. Each can has a different label identifying its contents: Memory Blocks, Data-Transmission Blocks, Interface Blocks. These are the parts needed for building up the chip.

Clear plastic tubes, carrying clear and tea-colored liquids, emerge from the mouths of the milk cans and drape across the table. The tubes end in fist-sized boxes mounted above shallow dishes sitting in a ring around the cans. As the different liquids drip into each dish, a beater like a kitchen mixer swirls the liquid. In each dish, nanomachines are building SuperChips.

A Nanofab "engineer," dressed in period clothing complete with name badge, is setting up a dish to begin building a new chip. "This," he says, holding up a blank with a pair of tweezers, "is a silicon chip like the ones made with pre-breakthrough technology.

Companies here in this valley made chips like these by melting silicon, freezing it into lumps, sawing the lumps into slices, polishing the slices, and then going through a long series of chemical and photographic steps. When they were done, they had a pattern of lines and blobs of different materials on the surface. Even the smallest of these blobs contained *billions* of atoms, and it took several blobs working together to store a single bit of information.

A chip this size, the size of our fingernail, could store only a fraction of a billion bits. Here at Nanofab, we used bare silicon chips as a base for building up nanomemory. The picture on the wall here shows the surface of a blank chip: no transistors, no memory circuits, just fine wires to connect up with the nanomemory we built on top. The nanomemory, even in the early days, stored *thousands* of billions of bits. And we made them like this, but a thousand at a time–" He places the chip in the dish, presses a button, and the dish begins to fill with liquid.

The chips in the dishes all look pretty much the same except for colour. The new chip looks like dull metal. The only difference we can see in the older chips, further along in the process, is a smooth rectangular patch covered by a film of darker material. An animated flowchart on the wall shows how layer upon layer of nanomemory building blocks arc grabbed from solution and laid down on the surface to make that film. The tourguide explains that the energy for this process, like the energy for molecular machines within cells, comes from dissolved chemicals–from oxygen and fuel molecules.

The total amount of energy needed here is trivial, because the amount of product is trivial: at the end of the process, the total thickness of nanomemory structure–the memory store for a Pocket Library–amounts to one-tenth the thickness of a sheet of paper, spread over an area smaller than a postage stamp.

Molecular Assembly

The animated flowchart showed nanomemory building blocks as big things containing about a hundred thousand atoms apiece (it takes a moment to remember that this is still submicroscopic). The build process in the dishes stacked these blocks to make the memory film on the SuperChip, but how were the blocks themselves built? The hard part in this molecular-manufacturing business has got to be at the bottom of the whole process, at the stage where molecules are put together to make large, complex parts.

The Silicon Valley Faire offers simulations of this molecular assembly process, and at no extra charge. From the tourguide, we learn that modern assembly processes are complex; that earlier processes–like those used by Nanofabricators, Inc. –used clever-but-obscure engineering tricks; and that the simplest, earliest concepts were never built. A short walk takes we to the Museum of Antique Concepts, the first wing of the Museum of Molecular Manufacturing.

A peek inside the first hall shows several people strolling around wearing loosely fitting jumpsuits with attached goggles and gloves, staring at nothing and playing mime with invisible objects. Stepping through the doorway while wearing the suit is entirely different.

The goggles show a normal world outside the door and a molecular world inside. Now we, too, can see and feel the exhibit that fills the hall. It's much like the earlier simulated molecular world: it shares the standard settings for size, strength, and speed. Again, atoms seem 40 million times larger, about the size of our fingertips. This simulation is a bit less thorough than the last was–we can feel simulated objects, but only with our gloved hands. Again, everything seems to be made of quivering masses of fused marbles, each an atom.

"Welcome," says the tourguide, "to a 1990 concept for a molecular-manufacturing plant. These exploratory engineering designs were never intended for actual use, yet they demonstrate the basics of molecular manufacturing: making parts, testing them, and assembling them."

Machinery fills the hall. Overall, the sight is reminiscent of an automated factory of the 1980s or 1990s. It seems clear enough what must be going on: Big machines stand beside a conveyor belt loaded with half-finished-looking blocks of some material; the machines must do some sort of work on the blocks. Judging by the conveyor belt, the blocks eventually move from one arm to the next until they turn a corner and enter the next hall.

Since nothing is real, the exhibit can't be damaged, so we walk up to a machine and give it a poke. It seems as solid as the wall of the nanocomputer. Suddenly, we notice something odd: no bombarding air molecules and no droplets of water–in fact, no loose molecules anywhere.

Every atom seems to be part of a mechanical system, quivering thermal vibration, but otherwise perfectly controlled. Everything here is like the nanocomputer or like the tough little gear; none of it resembles the loosely coiled protein or the roiling mass of the living cell.

The conveyor belt seems motionless. At regular intervals along the belt are blocks of material under construction: workpieces. The nearest block is about a hundred marble-bumps wide, so it must contain something like 100 x 100 x 100 atoms, a full million. This block looks strangely familiar, with its rods, crank, and the rest. It's a nanocomputer–or rather, a blocklike part of a nanocomputer still under construction.

Standing alongside the pieces of nanocomputer on the conveyor belt, dominating the hall, is a row of huge mechanisms. Their trunks rise from the floor, as thick as old oaks. Even though they bend over, they rear overhead. "Each machine," our tourguide says, "is the arm of a general-purpose molecular assembler.

One assembler arm is bent over with its tip pressed to a block on the conveyor belt. Walking closer, we see molecular assembly in action. The arm ends in a fist-sized knob with a few protruding marbles, like knuckles. Right now, two quivering marbles–atoms–are pressed into a small hollow in the block. As we watch, the two spheres shift, snapping into place in the block with a quick twitch of motion: a chemical reaction. The assembler arm just stands there, nearly motionless. The fist has lost two knuckles, and the block of nanocomputer is two atoms larger.

The tourguide holds forth: "This general-purpose assembler concept resembles, in essence, the factory robots of the 1980s. It is a computer-controlled mechanical arm that moves molecular tools according to a series of instructions. Each tool is like a single-shot stapler or rivet gun. It has a handle for the assembler to grab and comes loaded with a little bit of matter–a few atoms–which it attaches to the workpiece by a chemical reaction." This is like the rejoining of the protein chain in the earlier tour.

Molecular Precision

The atoms seemed to jump into place easily enough; can they jump out of place just as easily? By now the assembler arm has crept back from the surface, leaving a small gap, so we can reach in and poke at the newly added atoms. Poking and prying do no good: When we push as hard as we can (with our simulated fingers as strong as steel), the atoms don't budge by a visible amount. Strong molecular bonds hold them in place. Our pocket tourguide–which has been applying the power of a thousand 1990s supercomputers to the task of deciding when to speak up–remarks, "Molecular bonds hold things together.

In strong, stable materials atoms are either bonded, or they aren't, with no possibilities in between. Assemblers work by making and breaking bonds, so each step either succeeds perfectly or fails completely. In pre-breakthrough manufacturing, parts were always made and put together with small inaccuracies. These could add up to wreck product quality. At the molecular scale, these problems vanish. Since each step is perfectly precise, little errors can't add up. The process either works, or it doesn't."

But what about those definite, complete failures? Fired

by scientific curiosity, we walk to the next assembler, grab the tip, and shake it. Almost nothing happens. When we shove as hard as we can, the tip moves by about one-tenth of an atomic diameter, then springs back. "Thermal vibrations can cause mistakes by causing parts to come together and form bonds in the wrong place," the tourguide remarks. "Thermal vibrations make floppy objects bend further than stiff ones, and so these assembler arms were designed to be thick and stubby to make them very stiff. Error rates can be kept to one in a trillion, and so small products can be perfectly regular and perfectly identical.

Large products can be *almost* perfect, having just a few atoms out of place." This should mean high reliability. Oddly, most of the things we've been seeing outside have looked pretty ordinary–not slick, shiny, and perfect, but rough and homey. They must have been *manufactured* that way, or made by hand. Slick, shiny things must not impress anyone anymore.

Molecular Robotics

By now, the assembler arm has moved by several atom-widths. Through the translucent sides of the arm we can see that the arm is full of mechanisms: twirling shafts, gears, and large, slowly turning rings that drive the rotation and extension of joints along the trunk. The whole system is a huge, articulated robot arm. The arm is big because the smallest parts are the size of marbles, and the machinery inside that makes it move and bend has many, many parts. Inside, another mechanism is at work: The arm now ends in a hole, and we can see the old, spent molecular tool being retracted through a tube down the middle.

Patience, patience. Within a few minutes, a new tool is on its way back up the tube. Eventually, it reaches the end. Shafts twirl, gears turn, and clamps lock the tool in position. Other shafts twirl, and the arm slowly leans up against the workpiece again at a new site.

Finally, with a twitch of motion, more atoms jump across, and the block is again just a little bit bigger. The cycle begins again. This huge arm seems amazingly slow, but the standard simulation settings have shifted speeds by a factor of over 400 million. A few minutes of simulation time correspond to less than a millionth of a second of real time, so this stiff, sluggish arm is completing about a million operations per second.

Peering down at the very base of the assembler arm, we can get a glimpse of yet more assembler-arm machinery underneath the floor: Electric motors spin, and a nanocomputer chugs away, rods pumping furiously. All these rods and gears move quickly, sliding and turning many times for every cycle of the ponderous arm.

This seems inefficient; the mechanical vibrations must generate a lot of heat, so the electric motors must draw a lot of power. Having a computer control each arm is a lot more awkward now than it was in pre-breakthrough years. Back then, a robot arm was big and expensive and a computer was a cheap chip; now the computer is bigger than the arm. There must be a better way–but then, this is the Museum of Antique Concepts.

Building-Blocks into Buildings

Where do the blocks go, once the assemblers have finished with them? Following the conveyor belt past a dozen arms, we stroll to the end of the hall, turn the corner, and find ourself on a balcony overlooking a vaster hall beyond. Here, just off the conveyor belt, a block sits in a complex fixture. Its parts are moving, and an enormous arm looms over it like a construction crane.

After a moment, the tourguide speaks up and confirms our suspicion: "After manufacturing, each block is tested. Large arms pick up properly made blocks. In this hall, the larger arms assemble almost a thousand blocks of various kinds to make a complete nanocomputer.

The grand hall has its own conveyor belt, bearing a series of partially completed nanocomputers. Arrayed along this grand belt is a row of grand arms, able to swing to and fro, to reach down to lesser conveyor belts, pluck million-atom blocks from testing stations, and plug them into the grand workpieces, the nanocomputers under construction. The belt runs the length of the hall, and at the end, finished nanocomputers turn a corner–to a yet-grander hall beyond?

After gazing at the final-assembly hall for several minutes, we notice that nothing seems to have moved. Mere patience won't do: at the rate the smaller arms moved in the hall behind us, each block must take months to complete, and the grand block-handling arms are taking full advantage of the leisure this provides. Building a computer, start to finish, might take a terribly long time. Perhaps as long as the blink of an eye.

Molecular assemblers build blocks that go to block assemblers. The block assemblers build computers, which go to system assemblers, which build systems, which–at least one path from molecules to large products seems clear enough. If a car were assembled by normal-sized robots from a thousand pieces, each piece having been assembled by smaller robots from a thousand smaller pieces, and so on, down and down, then only ten levels of assembly process would separate cars from molecules.

Perhaps, around a few more corners and down a few more ever-larger halls, we would see a post-breakthrough car in the making, with unrecognizable engine parts and comfortable seating being snapped together in a century-long process in a hall so vast that the Pacific Ocean would be a puddle in the corner.

Just ten steps in size; eight, starting with blocks as big as the ones made in the hall behind us. The molecular world seems closer, viewed this way.

Molecular Processing

Stepping back into that hall, we wonder how the process begins. In every cycle of their sluggish motion, each molecular assembler gets a fresh tool through a tube from somewhere beneath the floor, and that *somewhere* is where the story of molecular precision begins. And so we ask, "Where do the tools come from?", and the tourguide replies, "We might want to take the elevator to our left."

Stepping out of the elevator and into the basement, we see a wide hall full of small conveyor belts and pulleys; a large pipe runs down the middle. A plaque on the wall says, "Mechanochemical processing concept, circa 1990." As usual, all the motions seem rather slow, but in this hall everything that seems designed to move is visibly in motion.

The general flow seems to be away from the pipe, through several steps, and then up through the ceiling toward the hall of assemblers above.

After walking over to the pipe, we can see that it is nearly transparent. Inside is a seething chaos of small molecules: the wall of the pipe is the boundary between loose molecules and controlled ones, but the loose molecules are well confined. In this simulation, our fingertips are like small molecules. No matter how hard we push, there's no way to drive our finger through the wall of the pipe.

Every few paces along the pipe a fitting juts out, a housing with a mechanically driven rotating thing, exposed to the liquid inside the pipe, but also exposed to a belt running over one of the pulleys, embedded in the housing. It's hard to see exactly what is happening.

The tourguide speaks up, saying, "Pockets on the rotor capture single molecules from the liquid in the pipe. Each rotor pocket has a size and shape that fits just one of the several different kinds of molecule in the liquid, so the process is rather selective. Captured molecules are then pushed into the pockets on the belt that's wrapped over the pulley there, then–"

"Enough," we say. Fine, it singles out molecules and sticks them into this maze of machinery. Presumably, the machines can sort the molecules to make sure the right kinds go to the right places.

The belts loop back and forth carrying big, knobby masses of molecules. Many of the pulleys–rollers?–press two belts together inside a housing with auxiliary rollers. While we are looking at one of these, the tourguide says, "Each knob on a belt is a mechanochemical-processing device.

When two knobs on different belts are pressed together in the right way, they are designed to transfer molecular fragments from one to another by means of a mechanically forced chemical reaction. In this way, small molecules are broken down, recombined, and finally joined to molecular tools of the sort used in the assemblers in the hall above. In this device here, the rollers create a pressure equal to the pressure found halfway to the centre of the Earth, speeding a reaction that–"

"Fine, fine," we say. Chemists in the old days managed to make amazingly complex molecules just by mixing different chemicals together in solution in the right order under the right conditions.

Here, molecules can certainly be brought together in the right order, and the conditions are much better controlled. It stands to reason that this carefully designed maze of pulleys and belts can do a better job of molecule processing than a test tube full of disorganized liquid ever could. From a liquid, through a sorter, into a mill, and out as tools: this seems to be the story of molecule processing. All the belts are loops, so the machinery just goes around and around, carrying and transforming molecular parts.

Beyond Antiques

This system of belts seems terribly simple and efficient, compared to the ponderous arms driven by frantic computers in the hall above. Why stop with making simple tools? We must have muttered this, because the tourguide speaks up again and says, "The Special-Assembler Exhibit shows another early molecular-manufacturing concept that uses the principles of this molecule-processing system to build large, complex objects. If a system is building only a single product, there is no need to have computers and flexible arms move parts around. It is far more efficient to build a machine in which everything just moves on belts at a constant speed, adding small parts to larger ones and then bringing the larger ones together as we saw at the end of the hall above."

This does seem like a more sensible way to churn out a lot of identical products, but it sounds like just more of the same. Gears like fused marbles, belts like coarse beadwork, drive shafts, pulleys, machines and more machines. In a few places, marbles snap into new patterns to prepare a tool or make a product. Roll, roll, chug, chug, pop, snap, then roll and chug some more.

Crude Technology

As the Silicon Valley Faire scenario shows, molecular manufacturing will work much like ordinary manufacturing, but with devices built so small that a single loose molecule of pollutant would be like a brick heaved into a machine tool. A leading company in computer-aided design, observes that nanotechnology and today's crude methods are very different: "Technology has never had this kind of precise control; all of our technologies today are bulk technologies. We take a big chunk of stuff and hack away at it until we're left with the object we want, or we assemble parts from components without regard to structure at the molecular level."

Molecular manufacturing will orchestrate atoms into products of symphonic complexity, but modern manufacturing mostly makes loud noises. These figurative noises are sometimes all too literal: A crack in a metal forging grows under stress, a wing fails, and a passenger jet crashes from the sky. A chemical reaction goes out of control, heat and pressure build, and a poisonous blast shakes the countryside. A lifesaving product cannot be made, a heart fails, and a hospital's heart-monitoring machine signals the end with a high-pitched wail.

Today, we make many things from metal, by machining. From the perspective of our standard, simulated molecular world, a typical metal part is a piece of terrain many days' journey across. The metal itself is weak compared to the bonds of the protein chain or other tough nanomechanisms: solid steel is no stronger than our simulated fingers, and the atoms on its surface can be pushed around with our bare hands. Standing on a piece of metal being machined in a lathe, we would see a cutting blade crawl past a few times per year, like a majestic plough the size of a mountain range.

Each pass would rip up a strip of the metal landscape, leaving a rugged valley broad enough to hold a town. This is machining from a nanotechnological perspective: a process that hacks crude shapes from intrinsically weak materials. Today, electronics are made from silicon chips. We have already seen the landscape of a finished chip. During manufacturing, metal features would be built up by a centuries-long drizzle of metal-atom rain, and hollows would be formed by a centuries-long submergence in an acid sea. From the perspective of our simulation, the whole process would resemble geology as much as manufacturing, with the slow layering of sedimentary deposits alternating with ages of erosion.

The term *nanotechnology* is sometimes used as a name for small-scale microtechnology, but the difference between molecular manufacturing and this sort of microlandscaping is like the difference between watchmaking and bulldozing.

Today, chemists make molecules by solution chemistry. We have seen what a liquid looks like in our first simulation, with molecules bumping and tumbling and wandering around. Just as assemblers can make chemical reactions occur by bringing molecules together mechanically, so reactions can occur when molecules bump at random through thermal vibration and motion in a liquid. Indeed, much of what we know today about chemical reactions comes from observing this process. Chemists make large molecules by mixing small molecules in a liquid. By choosing the right molecules and conditions, they can get a surprising measure of control over the results: only some pairs of molecules will react, and then only in certain ways.

Doing chemistry this way, though, is like trying to assemble a model car by putting the pieces in a box and shaking. This will only work with cleverly shaped pieces, and it is hard to make anything very complex. Chemists today consider it challenging to make a precise, three-dimensional structure having a hundred atoms, and making one with a thousand atoms is a great accomplishment.

Molecular manufacturing, in contrast, will routinely assemble millions or billions. The basic chemical principles will be the same, but control and reliability will be vastly greater. It is the difference between throwing things together blindly and putting them together with a watchmaker's care. Technology today doesn't permit thorough control of the structure of matter. Molecular manufacturing will. Today's technologies have given us computers, spacecraft, indoor plumbing, and the other wonders of the modern age. Tomorrow's will do much more, bringing change and choices.

Simple Matter, Smart Matter

Today's technology mostly works with matter in a few basic forms: gases, liquids, and solids. Though each form has many varieties, all are comparatively simple.

Gases, as we've seen, consist of molecules ricocheting through space. A volume of gas will push against its walls and, if not walled in, expand without limit. Gases can supply certain raw materials for nanomachines, and nanomachines can be used to remove pollutants from air and turn them into something else. Gases lack structure, so they will remain simple.

Liquids are somewhat like gases, but their molecules cling together to form a coherent blob that won't expand beyond a certain limit. Liquids will be good sources of raw materials for nanomachines because they are denser and can carry a wide range of fuels and raw materials in solution (the pipe in the molecular-processing hall contained liquid). Nanomachines can clean up polluted water as easily as air, removing and transforming noxious molecules. Liquids have more structure than gases, but nanotechnology will have its greatest application to solids.

Solids are diverse. Solid butter consists of molecules stronger than steel, but the molecules cling to one another by the weaker forces of molecular stickiness. A little heat increases thermal vibrations and makes the solid structure disintegrate into a blob of liquid. Butterlike materials would make poor nanomachines.

Metals consist of atoms held together by stronger forces, and so they can be structurally stronger and able to withstand higher temperatures. The forces are not very directional, though, and so planes of metal atoms can slip past one another under pressure; this is why spoons bend, rather than break. This ability to slip makes metals less brittle and easier to shape (with crude technology), but it also weakens them. Only the strongest, hardest, highest-melting-point metals are worth considering as parts of nanomachines.

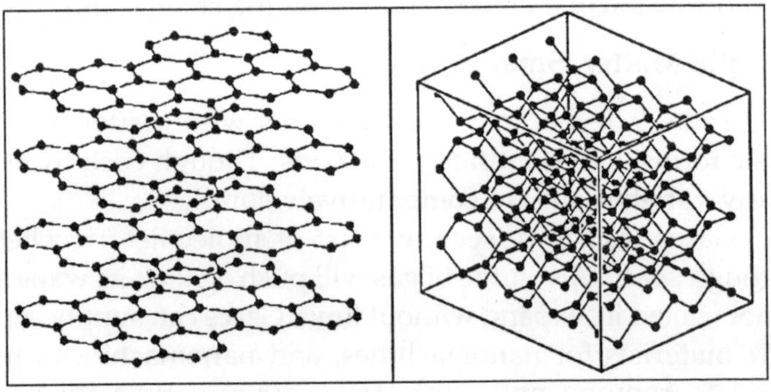

Fig. Carbon-Soft And Hard

On the left is graphite–the material called "lead" in pencils–made of carbon atoms. On right is diamond–the same atoms arranged in a different pattern.

Diamond consists of carbon atoms held together by strong, directional bonds, like the bonds down the axis of a protein chain. These directional bonds make it hard for planes of atoms to slip past one another, making diamond (and similar materials) very strong indeed–ten to a hundred times stronger than steel. But the planes can't easily slip, so when the material fails, it doesn't bend, it breaks. Tiny cracks can easily grow, making a large object seem weak. Glass is a similar material: glass windows seem weak–and a scratch makes glass far weaker–yet thin, perfect glass fibers are widely used to make composite materials stronger and lighter than steel. Nanotechnology will be able to build with diamond and similar strong materials, making small, flawless fibers and components.

In engineering today, diamond is just beginning to be used. Japan has pioneered a technology for making diamond at low pressure, and a Japanese company sells a speaker with excellent high-frequency response–the speaker cone is reinforced with a light, stiff film of diamond. Diamond is extraordinary stuff, made from cheap materials like natural gas. U.S. companies are scrambling to catch up.

All these materials are simple. More complex structures lead to more complex properties, and begin to give some hint of what molecular manufacturing will mean for materials.

What if we strung carbon atoms in long chains with side-groups, a bit like a protein chain, and linked them into a big three-dimensional mesh? If the chains were kinked so that they couldn't pack tightly, they would coil up and flop around almost like molecules in a liquid, yet the strong bonds would keep the overall mesh intact. Pulling the whole network would tend to straighten the chains, but their writhing motions would tend to coil them back up. This sort of network has been made: it is called rubber.

Rubber is weak mostly because the network is irregular. When stretched, first one chain breaks, then another, because they don't all become taut at the same time to share and divide the load. A more regular mesh would be as soft as rubber at first, but when stretched to the limit would become stronger than steel. Molecular manufacturing could make such stuff.

The natural world contains a host of good materials–cellulose and lignin in wood, stronger-than-steel proteins in spider's silk, hard ceramics in grains of sand, and more.

Many products of molecular manufacturing will be designed for great durability, like sand. Others will be designed to fall apart easily for easy recycling, like wood. Some may be designed for uses where they may be thrown away. In this last category, nanotailored biodegradables will shine. With care, almost any sort of product from a shoe to computer-driven nanomachines can be made to last for a good long time, and then unzip fairly rapidly and very thoroughly into molecules and other bits of stuff all of kinds normally found in the soil.

This gives only a hint of what molecular manufacturing will make possible by giving better control of the structure of solid matter. The most impressive applications will not be superstrong structural materials, improved rubber, and simple biodegradable materials: these are uniform, repetitive structures not greatly different from ordinary materials. These materials are "stupid." When pushed, they resist, or they stretch and bounce back. If we shine light on them, they transmit it, reflect it, or absorb it. But molecular manufacturing can do much more. Rather than heaping up simple molecules, it can build materials from trillions of motors, ratchets, light-emitters, and computers.

Muscle is smarter than rubber because it contains molecular machines: it can be told to contract. The products of molecular manufacturing can include materials able to change shape, colour, and other properties on command. When a dust mote can contain a supercomputer, materials can be made smart, medicine can be made sophisticated, and the world will be a different place.

IDEAS AND CRITICISMS

We've just seen a picture of molecular manufacturing (of one sort) and of what it can do (in sketchy outline). Now let's look at the idea of nanotechnology itself: Where did it come from, and what do the experts think of it?

Origins

The idea of molecular nanotechnology, like most ideas, has roots stretching far back in time. In ancient Greece, Democritus suggested that the world was built of durable, invisible particles–atoms, the building blocks of solid objects, liquids, and gases. In the last hundred years, scientists have learned more and more about these building blocks, and chemists have learned more and more ways to combine them to make new things. Decades ago, biologists found molecules that do complex things; they termed them "molecular machines."

Physicist Richard Feynman was a visionary of miniaturization who pointed toward something like molecular nanotechnology: on December 29, 1959, in an after-dinner talk at the annual meeting of the American Physical Society, he proposed that large machines could be used to make smaller machines, which could make still smaller ones, working in a top-down fashion from the macroscale to the microscale. At the end of his talk, he painted a vision of moving individual atoms, pointing out, "The principles of physics, as far as I can see, do not speak against the possibility of maneuvering things atom by atom."

He pictured making molecules, pointing clearly in the direction taken by the modern concept of nanotechnology: "But it is interesting that it would be, in principle, possible (I think) for a physicist to synthesize any chemical substance the chemist writes down. Give the orders, and the physicist synthesizes it. How? Put the atoms down where the chemist says, and so we make the substance."

Despite this clear signpost pointing to a potentially revolutionary area, no one filled the conceptual gap between miniature machines and chemical substances. There was no clear concept of making molecular machines able to build more such machines, no notion of controllable molecular manufacturing.

With hindsight, one wonders why the gap took so long to fill. Feynman himself didn't follow it up, saying that the ability to maneuver atoms one by one "will really be useless" since chemists would come up with traditional, bulk-process ways to make new chemical substances. For a researcher whose main interest was physics, he had contributed much just by placing the signpost: it was up to others to move forward. Instead, the idea of molecular machines for molecular manufacturing didn't appear for decades.

From today's viewpoint, molecular nanotechnology looks more like an extension of chemistry than like an extension of miniaturization. A mechanical engineer, looking at nanotechnology, might ask, "How can machines be made so small?" A chemist, though, would ask, "How can molecules be made so large?" The chemist has the better question. Nanotechnology isn't primarily about miniaturizing machines, but about extending precise control of molecular structure to larger and larger scales. Nanotechnology is about making (precise) things *big*.

Table: Macroscopic and Molecular Components

Technology	*Function*	*Molecular Examples*
Struts, beams, casins	Transmit force, hold positions	Cell walls, microtubules
Cables	Transmit tension	Collagen, Silk
Fasteners, glue	Connect parts	Intermolecular forces
Solenoids, actuators	Move things	Muscle actin, myosin
Motors	Turn shafts	Flagellar motor
Drive shafts	Transmit torque	Bacterial flagella
Bearings	Support moving parts	Single bonds
Clamps	Hold workpieces sites	Enzymatic binding
Tools	Modify workpieces molecules	Enzymes, reactive
Production lines	Control devices ribosomes	Enzyme systems,
Numerical control systems	Store and read programs	Genetic system

Nature gives the most obvious clues to how this can be done, and it was the growing scientific literature on natural molecular machines that led one of the present authors (Drexler) to propose molecular nanotechnology of the sort described here. A strategy to reach the goal was part of the concept: Build increasingly complex molecular machinery from simpler pieces, including molecular machines able to build more molecular machines. And the motivation for studying this, and publishing? Largely the fear of living in a world that might rush into the new technology blindly, with ugly consequences.

Life might be much simpler if these ideas about nanotechnology had some fatal flaw. If only molecules couldn't be used to form machines, or the machines couldn't be used to build things, then we might be able to keep right on going with our crude technologies: our medicine that doesn't heal, our spacecraft that don't open a new frontier, our oil crises, our pollution, and all the limits that keep us from trading familiar problems for strange ones. Most new ideas are wrong, especially if they purport to bring radical changes. It is not unreasonable to hope that these are wrong. From years of discussions with chemists, physicists, and engineers, it is possible to compile what seems to be a complete list of basic, critical questions about whether nanotechnology will work. The questioners generally seem satisfied with the answers.

The earlier scenarios describe the nature of thermal vibration and the problems it can cause. Designing nanomachines strong enough and stiff enough to operate reliably despite thermal vibration is a genuine engineering challenge. But calculating the design requirements usually requires only simple textbook principles, and these requirements can be met for everything described in this book.

Quantum mechanics says that particles must be described as small smears of probability, not as points with perfectly defined locations. This is, in fact, why the atoms and molecules in the simulations felt so soft and smooth: their electrons are smeared out over the whole volume of the molecule, and these electron clouds taper off smoothly and softly toward the edges. Atoms themselves are a bit uncertain in position, but this is a small effect compared to thermal vibrations. Again, simple textbook principles apply, and well-designed molecular machines will work.

Chemists work with loose molecules in liquids, and they naturally tend to picture molecules as flying around loose. It is possible to build nanomachines and molecular-manufacturing systems that work in this sort of environment (biological mechanisms are an existence proof), but in the long run, there will be no need to do so. The Silicon Valley Faire simulation gives the right idea: Systems can be built with no loose molecules, making nanomechanical design much easier. If no molecules are loose inside a machine, then loose molecules can't cause problems there.

Chemists perform chemical reactions, which means that they tend to work with reactive, unstable molecules. Many molecules, though, can sit around in peace with their neighbors for millions of years, as is known both from chemical theory and from the study of molecules trapped in ancient rock. Nanomachines can be built from the more stable sorts of structure. The only necessary exception is in molecular assembly, where molecules must react, but even here the reactive molecules need not be turned loose. They can be applied just when and where they are needed in the construction process.

An easy way to explain molecular manufacturing is to say that it is somewhat like molecular biology: small, complex molecular devices working together to build things and do various jobs. The next point, however, is that molecular manufacturing is different in every detail and different in overall structure: compare the nanocomputers, assembler arms, and conveyor belts described above to the shaggy. Biology is complex in a strange and wonderful way. Engineers need not even understand life, much less duplicate it, merely to build a molecular-scale factory.

Nanotechnology is basically a shotgun marriage of chemistry and mechanical engineering, with physics (as always) presiding. This makes a complete evaluation difficult for most of today's specialists, because each of these fields is taught separately and usually practiced separately. Many specialists, having highly focused backgrounds, find themselves unequipped to evaluate proposals that overlap other disciplines. When asked to do so, they will state feelings of discomfort, because although they can't identify any particular problems, they can't verify the entire concept as sound. Scientists and engineers with multidisciplinary backgrounds, or with access to specialists from other fields, can evaluate the idea from all sides.

When physicists, chemists, biologists, engineers, and computer scientists evaluate those parts of nanotechnology that fall within their disciplines, they agree: At no point would it require new principles or violate a physical law. There may for many years be some experts offering negative off-the-cuff opinions, but the consensus among those who have taken the time to examine the facts is clear. Molecular nanotechnology falls entirely within the realm of the possible.

Increased human abilities have routinely been used to damage the environment and to make war. Even the crude technologies of the twentieth century have taken us to the brink. It is natural to feel exhilarated (or terrified) by a prospect that promises (or threatens) to extend human abilities beyond most past dreams (or nightmares). It is better to feel both, to meld and moderate these feelings, and to set out on a course of action that makes bad outcomes less likely. We're convinced that the best course is to focus on the potential good while warning of the potential evils.

Those in failing health may be justified in saying this; others are expressing an opinion that may well be wrong. It would be optimistic to assume that benefits are around the corner, and prudent to assume that they will be long delayed. Conversely, it would be optimistic to assume that dangers will be long delayed, and prudent to assume that they will arrive promptly. Whatever good or ill may come of post-breakthrough capabilities, the turbulence of the coming transition will present a real danger. While we invite readers to take a "What if?" stance toward these technologies, it would be imprudent to listen to the lulling sound of the promise "not in our lifetimes." Even today, public acceptance of man's coming exploration of space is slow. It is considered an event our children may experience, but certainly not one that we shall see.

Perspective

We are still many years away from nanotechnology based on molecular manufacturing. It might even seem that such vast, slow giants as ourselves could never make such small, quick machines. We'll try to get some feel for the road ahead, for its length, and for how fast we're moving.

We are already surprisingly close to developing a crude molecular manufacturing technology, and getting visibly closer every week. The first, crude technology will enable the construction of molecular machines that can be used to build better molecular machines, climbing a ladder of capabilities that leads to general-purpose molecular assemblers as good or better than those described here.

Chapter 3

Micro Production Technology

In the last years the demand for ultra-precise production facilities increased rapidly. The revolution in MEMS technology and in communication technology leads to a variety of small objects that have to be assembled and bonded with nano-precision. On the way to new handing and assembly systems a set of tasks has to be fulfilled:

Fig. The Nanomotor.

Task 1: Reproducibility of positioners: better than 1 ìm

Task 2: Adjustment of the microparts orientation: better than 20 nm

Task 3: Many degrees of freedom in smallest volume

Task 4: Compatible system of modules

Task 5: Handing over from coarse approach to microassembly
Task 6: Protection of sensitive microparts and grippers (force limitations)
Task 7: Transport "without" gravity
Task 8: Integrated sensors
Task 9: Quality control with suitable resolution
Task 10: Micro adhesive bonding technology.

As an actuator for microtechnology the Nanomotor developed by Klocke Nanotechnik is a central element that allows to solve most of the items listed. It is a linear motor with a positioning stroke of up to some centimeters at atomic resolution. This is a bridge over eight orders of magnitude. The small version of the Nanomotor is half the size of a matchstick and can lift six times its own mass or make indents even into a diamond.

The Nanomotor can be combined to many kinds of positioners: From scanning tunneling microscopes with atomic resolution on a normal table over positioning tables and manipulators up to microassembly stages with different kind of grippers and glue dispensers. It is the driving element for the Nanorobotics series presented below.

REPRODUCIBILITY

The assembly line for mobile phone production has a typical movement reproducibility of 20 ìm. Special assembly lines are offered with a guaranteed resolution of 10 ìm. These systems can be tuned to a reproducibility of about 5 ìm by using granite plates at constant temperatures. For glass fibre assembly the demand in reproducibility is 3 ìm at the moment, a value that may be achievable. Next generations will need 1 ìm and a future-proof system should allow reproducibility below 1 ìm. But thermal expansions in the classical design of meter sized robots and actuators will already avoid a precise positioning.

The reproducibility of positioners enclosing Nanomotors is better than 50 nm. The resolution is even much better: atomic without damping systems.

Fig. Atomic Structures of a Gold Surface

The picture was taken at a fair with a Scanning Tunneling Microscope (STM) without any vibration damping system. Using a Nanomotor as central element reduces the size and makes the STM insensitive to vibrations.

ORIENTATION OF MICROPARTS

A good reproducibility is only the first step. At the moment when the micropart shall be fixed it is often necessary to adjust its orientation with a resolution far below 1 ìm. Sometimes optical fibers have to be aligned with a precision better 10 nm.

Sharpen a hair or bring a tip onto the end of a glass fibre and focus it! This microassembly example of a detection element in Scanning Nearfield Optical Microscopes (SNOM) needs a resolution of better 20 nm^2. A pyramid-shaped microtip is glued onto the end of a glass fibre.

A 150 nm small hole in that tip forms a lens as sub-ë light source. The adhesive bonding process requires a very high precision. During the adjustment and the following curing of the adhesive, laser light is coupled through the glass fibre. The light emission of the microtip is measured to control the orientation of the microtip and glass fibre continuously. A misalignment of only 30 nm is clearly visible. The glass fibre with such a pinhole at its end then is assembled beside a shear force sensor. This device is used as central detection element in SNOMs.

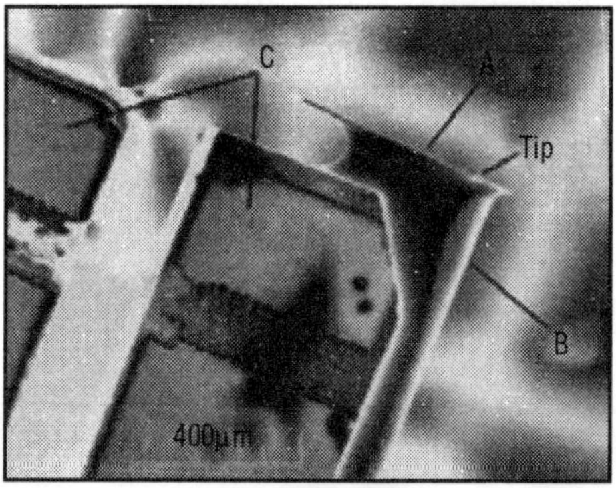

Fig. Assembled SNOM tip

- Cantilever with microtip
- Glass fibre
- Shear force sensor

The amount of adhésive applied between the parts a, b and c is critical: too much adhesive and the shear force sensor is damped. Too little adhesive allows for the cantilever of the microtip to oscillate freely. In both cases the device would be destroyed. The adhesive is applied with one of the Nanorobotics manipulators.

MANY DEGREES OF FREEDOM IN SMALLEST VOLUME

When the size of a positioning system is doubled, a factor of 16 more acoustic noise will couple into it. A small size is essential to achieve a good resolution without expensive damping systems. For many assembly and handling tasks it is necessary, that a set of different sensors and actuators can move to one point. This point is defined e.g. by the focus of three or more video microscope cameras. The smaller the actuators are the more of them can operate in this area.

COMPATIBLE SYSTEM OF MODULES

For assembly lines, the easy exchange of compatible

positioners is necessary to reduce down time. Flexibility is also necessary, when new MEMS products are developed. During development, the assembly stage can vary often in strokes and degrees of freedom. Positioners, microgrippers, adhesive dispensers, or e.g. nano-indenters should be screwed onto each other within some minutes, all controlled with the same network compatible electronics and software.

In the last years Klocke Nanotechnik and partners developed a new Nanorobotics system. Loads of up to two kilogram can be moved over strokes up to 70 mm at resolutions down to 2 nm.

Fig. (a) Linear stage with 10 mm stroke at 2 nm resolution, burdened with 2 kg. (b) A set microgrippers with differenttips. Resolution of movement: Atomic. (c) A small manipulators with 5 × 15 × 20 mm^3 stroke

This system allows microassembly, interconnection technology, analysis and reliability testing with plenty degrees of freedom in smallest volume without backlash. All Nanorobotics modules can easily be fixed onto each other. Equipped with high resolution position sensors they allow automation with a repeatability far below 50 nm. The following pictures show examples of the Nanorobotics modules.

The combination of a manipulator like shown in Figure and a microgripper is already a small microassembly stage. The primary objective during the development of these Nanorobotics modules was a resolution in movement of a few single Nanometers.

This is made possible in principle by the Nanomotor: Compared with other high stroke piezo drives the Nanomotor has a real fine positioning stroke of more than a micron. Within this stroke the Nanomotor behaves exactly like a piezo, without backlash. The combination of the Nanomotor with mechanical guiding elements like ball bearing tables or gears should also eliminate the play of the mechanical guiding element and lead to a superior movement resolution, free of any backlash. But it didn't.

The movement of a first prototype showed a kind of backlash with 50 nm amplitude in an Electron Microscope. This value is much smaller than mechanical play and too big for a piezo drive. The analysis of the whole mechanical chain in this linear stage with a high resolution Electron Microscope led to the source of this backlash. The movement of this small motorized stage caused a compression of materials used in this setup. Even the elastic constants of materials reduce the performance of this device.

After this result the linear stage was developed in a new design with other materials and geometries. The backlash caused by compression could be reduced from 50 nm down to an amplitude of only 2 nm. This is the distance of 10 atoms! Furthermore, compared with other piezo drives the Nanomotor does not press in this design against the mechanical guiding elements from the side and so avoids additional off-axis movements.

The design was defined for linear stages as well as for rotary devices or grippers. Only with this precision the Task 1 and Task 2 could be fulfilled in one system. These Nanorobotics modules are the key elements for the precision of the Micro Production System described. Also for manipulation tasks in Electron Microscopes the achieved resolution is essential.

HANDING OVER FROM COARSE APPROACH TO MICROASSEMBLY

A glass fibre is a typical object that has to be assembled with (sub-) micron resolution, but comes from a cable drum, perhaps in a machine hall. The robot supplying the glass fibre stops its movement not exactly in the zero position of the microgripper. Pattern recognition systems may help, but they are expensive. The Nanomotor driven microgrippers can solve the problem of handing over from coarse to good accuracy. The gripper structure can grip a fibre even when it is out of the grippers centre: Both tips close until one is stopped by the fibre. The second tip continues its movement until the gripper is closed. Then the fibre is released (e.g. by the robot) and the gripper structure moves into its own centre position.

PROTECTION OF MICROPARTS AND GRIPPERS

The combination of microassembly stages and microgrippers with classical CNC technology includes miscellaneous risks that may be new for both sides. It is possible to fix a microgripper onto a CNC robot. But what happens, if this robot drives the microgripper a few mm too deep into the table? It happens in practice, for example during setting-up the operation.

The microassembly stages made with the Nanorobotics modules are inherently protected against a crash. Each axis posesses a force limitation and the microgripper is designed compatible to this force limit. The gripper can be driven down onto a base plate without damaging its small tips or its bending elements.

Horizontal movements slightly hitting the gripper structure against a small wall also do not harm the gripper. The protection of microparts is important as well. A task that required superior sensitivity and manipulation skills was solved at KOSMA. The assembly of substrate bars for RF devices with a profile of 35 ìm x 80 ìm could only be done with Nanometer resolution, because coarser tools would have damaged or flipped away the components. The sensitivity of the substrate bars for electrostatic discharge is much higher than the sensitivity of semiconductor devices. The small size of the assembly stage allowed an easy grounding and a secure handling. This assembly stage is now used by RF groups from all over the world to produce high frequency devices (800 GHz or higher).

TRANSPORT "WITHOUT" GRAVITY

When the size of microparts is smaller than the edge

sharpness of a microgripper, other methods of transport should be used. Then, the influence of gravity onto the objects to be handled can be smaller than other forces which in turn can be expolited for the transport. The developments of such processes and tools were made in an electron microscope. The enhancement of the electron microscope from the pure device for analytics to a material processing system was enabled with the Nanomanipulator.

Small as a film cartridge even several manipulators fit into the vacuum chamber. The manipulator can be moved half a cubic centimeter with a resolution of one Nanometer. It can contain various micro tools for cutting, scraping, gripping, spark erosion or for laser ablation. The combination of Nanomanipulator and SEM leads to a powerful tool as a workbench. It allows manipulation and even assembly with Nanometer resolution in combination with all kinds of analytical methods (EDX, WDX, EBSP).

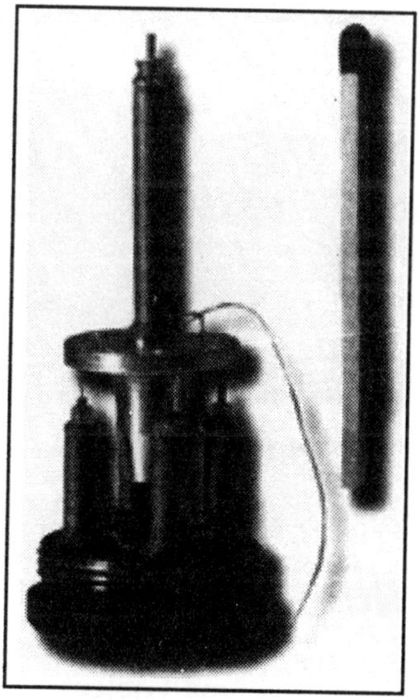

Fig. The Nanomanipulator

The sequence is an example of handling 1 ìm small objects in a SEM. The left manipulator lifts up a cluster of these balls with the first force. The second manipulator captures this cluster with a second higher force. In this case the first force was adhesion, the second magnetism.

Fig. Nanomanipulation of Magnetic Beads

In the handling process another force is used. Electronical drop outs in integrated circuits that result from structural defects can be inspected by high-resolution transmission electron microscopy measurements. This requires to cut a very small lamella out of the drop-out-area.

Using Nanomanipulators renders the complicated preparation techniques much easier. First the Nanomanipulator breaks a sample out of a chip structure and then a voltage is applied only to the tip of the Nanomanipulator. This additional charge allows to grip the sample and raise it.

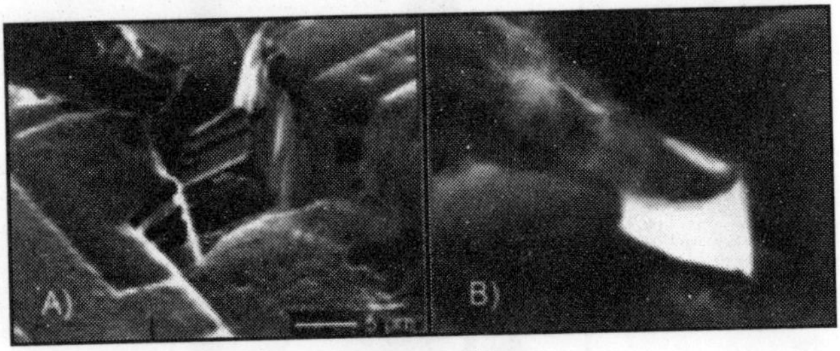

INTEGRATED SENSORS

Additional sensors can help during assembly. Besides vision control systems force sensors are important. Imagine a tool with 2 nanometer resolution that is controlled by a joystick and answers with the corresponding force feedback.

Then, manipulating small objects manually becomes very intuitive, because the operator uses his own senses for controlling the stage. The forces applied to the tool tip can be so high, that even imprints in diamond surfaces can be produced.

The force sensor between manipulator and sample allows to apply or measure calibrated forces in mN range by moving the actuator within some 100 nm. Values like friction forces, adhesive stiffness, breakage limits or elastic constants of microparts can be determined in high resolution.

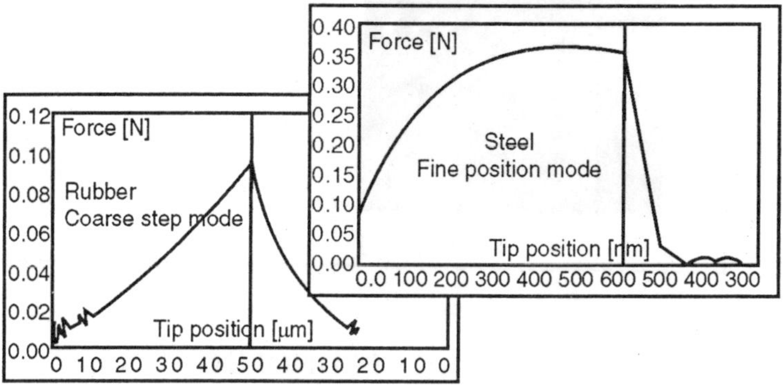

Fig. Force-Stroke Diagram of (a) Rubber and (b) Steel

The force sensor can also be assembled between a z-axis and a microgripper. Calibrated snap-in forces can then be applied, e.g. to press glass fibers into spring structures.

SUITABLE QUALITY CONTROL

In classical precision mechanics the device size is about 1 m and the precision is up to 1 ìm - a ratio of 6 orders of magnitude. A micropart produced with 0.05 mm size leads to 50 Picometers as comparable precision, one tenth of the radius of a gold atom. Instruments to measure the precision of microparts and the alignment of them should have Nanometer resolution. Such ultra precise quality control systems like the MicroProf from FRT are integrated in the Micro Production System.

Some quality control instruments have to face an additional problem: an imprint to measure the hardness of a sample must be much smaller than the micropart that shall be analyzed. But when such a small imprint is made and the micropart is transferred to the analysis instrument that shall measure the depth of that imprint - where is it? If microassembly and quality control can be made in one absolute positioning system many problems are solved at once.

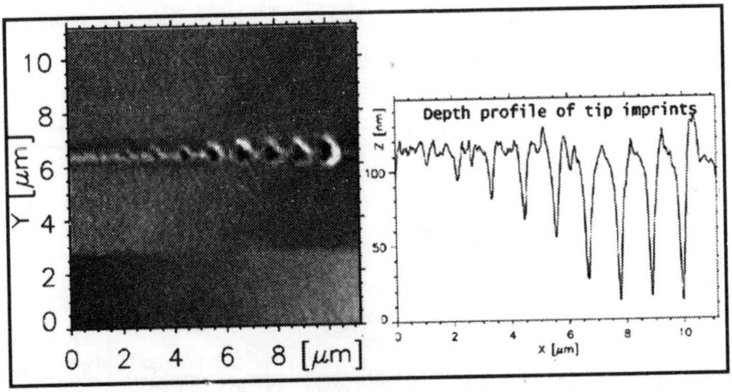

Fig. Tip imprints made in ruby with a Nanomotor

ADHESIVE BONDING TECHNOLOGY ON THE MICRO-SCALE

In many instances, adhesive bonding is the only bonding technology feasible in the micro world. This is due to the small size of the parts to be joined and to the great variety of very specialized materials. Each of these materials has its specific requirements for the bonding technology which oftentimes can only be simultaneously fulfilled by the adhesive bonding technology. Fortunately, adhesive bonding can offer a couple of advantages:

- Joining of many, dissimilar materials
- Multi-functionality of the joint (e.g. mechanical fastening plus conductivity)
- Low heat/ cold joining
- Small mechanical stress
- Even stress distribution
- Galvanic isolation of the parts (no contact corrosion)
- New freedom for design
- Innovative technical solutions

The development of an adhesive bonding process consists of several steps:

- Specification of the end-product and deducted

specification of the adhesive joint, e.g. mechanical, thermal, electrical stress, media, precision, optical properties, service life.

- *Selection of the parts to be joined (material, surface):* The chemical nature and topography of the surface effect the choice of adhesives and the necessity of a surface pre-treatment prior to bonding. The temperature resistance and the transmission characteristics impact the possibilities for thermal or radiation curing.
- *Geometrical construction/ design:* There are general guidelines for optimal design of adhesive joints. However, they must be harmonized with other requirements, e.g. stemming from the production process or regarding the functionality of the system.
- *Requirements of the production:* The host of the respective aspects to be considered include degree of automation, prevalent equipment for mixing, dosing, curing adhesives, on-line inspection/ monitoring, precision, environment, subsequent production steps.
- *Selection of adhesive:* Characteristics to be considered are state of the adhesive as delivered (paste, foil, no mix, 2 component...), shelf life, pot life, reactive/ hotmelt/combination, curing mechanism(s), rheology (=> mixing, dosing, application technology, wetting), requirements for the environment (e.g. humidity), adhesion (surface pre-treatment/ primer necessary?), properties after curing (mechanical, optical, chemical...), repair/recycling.
- *Micro assembly:* A discussion of the aspects of micro assembly can be found throughout the other parts of the paper. The micro assembly device has to cater for the requirements of the adhesive bonding process such as a defined gap of the adhesive bondline, or a defined force for placing the component to be bonded.
- *Cure*: There are different curing mechanisms available, which can be tailored to the task. A two

stage curing allows for a fast first cure – e.g. for fixing a precision alignment – followed by a second, slow, final curing step.

- *Quality Control:* The Quality Control can encompass the control of the adhesive as delivered (e.g. chemical reactivity or chemical groups present), the on-line inspection of crucial steps (e.g. application of the adhesive or precision of the assembly before cure), or the quality of the final joint.

Properties of Adhesives in Small Dimensions

The particularity of adhesives in small dimensions is their large surface and their small volume. Before curing, this means that there is a great interaction between the adhesive and the surrounding atmosphere.

This can render the adhesive sensitive to the atmosphere's composition (e.g. humidity or oxygen) and it can change the adhesives constitution.

Thus, the adhesives curing behaviour or its properties after curing can be altered compared to bulk adhesives. As a consequence, adhesives which may be suitable for a similar task, however, on a macroscopic scale can be utterly unsuitable on the microscopic scale. Therefore, the development of an adhesive process has by all means to take the geometric size of the adhesive into account and according investigations have to be undertaken. The well known finite element calculation tools can be applied to calculate the bond properties. However, the material laws and data must be true for the small dimensions.

Adhesive bonding quite often is the only feasible joining technology on the micro scale, especially when dissimilar materials are to be bonded. At the same time, adhesive bonding can offer substantial technical advantages and allow for novel designs.

Many aspects of the Micro Production System were developed with respect to adhesive bonding procedures. A team of experts in this new technology offers the development of micro-manufacturing steps as a service in combination with this Micro Production System. It includes Nanorobotics modules, microassembly, micro adhesive bonding, and quality control in one absolute positioning system.

Fig. Dispensed Micro-Dots of an Electrically Conductive Adhesive

MICRO PRODUCTION SYSTEM

The Micro Production System developed by Klocke Nanotechnik and partners bridges the gap between nanotechnology and the classical mechanical engineering. It combines the advantages of both technologies: the backlash free movement with Nanometer resolution and the capability of up to kilograms of load at huge strokes.

The included Nanorobotics modules offer plenty degrees of movement and a repeatability of better 50 nm, a value far away from the Micrometer values of classical production systems. The Micro Production System is available in a lot of different levels of complexity. It was the main goal during the development of this system to allow for flexibility:

- Modular design, to be able to add, combine, or exchange components such as force sensors, wafer prober, manipulators, vacuum- and microgrippers or Scanning Probe Microscopes in order to produce different products.
- Ability to scale-up the production beginning from first manually assembled samples up to fully automated mass production.
- Process control with an easy to use software.

An example shall detail the second option. The first step

to a product is a prototype that has to be assembled, e.g. with a 4 axis microassembly stage (X, Y, Z-stages plus microgripper) and a joystick.

Then about 10-100 prototypes are necessary and some assembly steps should run automatically, e.g. by integrated position sensors and a small software packet. An adhesive dispenser may be added, perhaps also an electrical prober or a Nanofinger.

In the next step a first production series of 1000-5000 pieces needs more assistance and closed loop operation. Some video cameras and a vision system help to identify microparts, a base stage and rotary drives arrange these microparts for the gripper and a sequencer software allows to programme complete assembly processes just by "drag and drop". When prototypes turn into products quality control gets more important and the system can be expanded for example with Scanning Probe Microscopes, nano-indenters, surface inspection systems or wafer probers.

The compatible modules of the Nanorobotics series allows this scale-up from a single microgripper over small microassembly stages up to a Micro Production Systems with plenty of different configurations – always with nanoresolution. In the same way the Ethernet-based electronics and modular software can grow with the system. This concept is the result of our experience in our Microtechnology network and with a lot of companies who have to overcome the scale-up problem.

DIFFERENT HARDWARE ARRANGEMENTS

Starting form a microassembly stage made with Nanorobotics modules like shown in Figure the system can be expanded e.g. by an additional manipulator with an adhesive dispenser or some more modules to move additional devices like a vacuum gripper or a wafer prober.

When a bigger high precision xy-stage is necessary there are two systems defined by the size of this stage: an xy-stage with 100x100 mm^2 stroke or a big version with 350x350 mm^2 stroke. Both stages have a resolution better 200 nm.

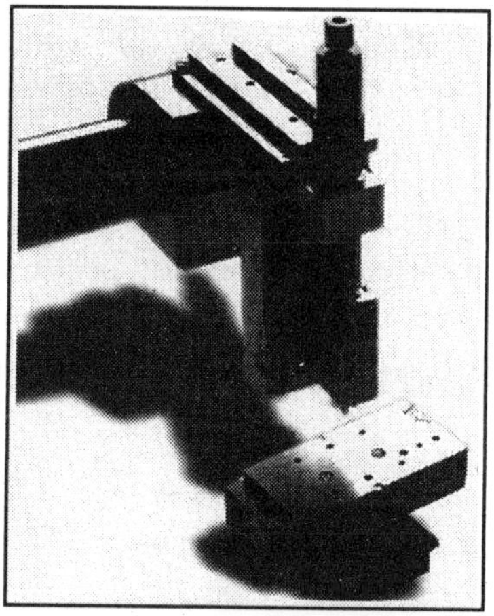

Fig. Microassembly Stage made with Nanorobotics Modules, 20 mm stroke each

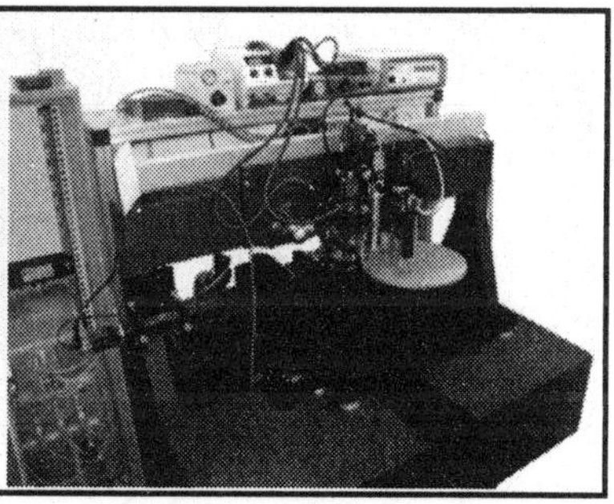

Fig. Micro Production system with Small xy-stage

Fig. Micro Production System with a Bigger xy-stage

The different stroke of the two xy-stages is coupled with a different size of the whole system. In the small system only one rotary drive fits onto the xy-stage. Source and target area have to fit onto this rotary drive. The bridge over the bottom area is big enough to include a set of Nanorobotics modules that can operate with 5 independent tools. Three video microscope cameras can be fixed at the frame structure.

In comparison with this setup the bigger xy-stage offers much more space for equipment. Two rotary drives can be moved with the bottom xy-stage, one for the source area, the other for the target area. Besides the Nanorobotics modules of the smaller setup many additional components like adhesive dispenser, surface inspection systems or wafer prober can be included.

One video camera is fixed on the moving part of the xy-stage for a programmable inspection position. This big system can include up to 6 video cameras.

COMPONENTS OF THE MICRO PRODUCTION SYSTEM

The Micro Production System includes different groups of positioners, actuators and sensors that are summarized in Figure.

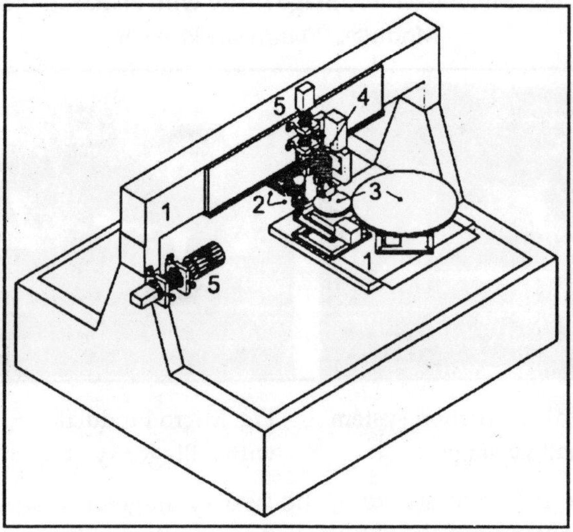

Fig. Scheme of the Micro Production System

- High precision xy-stage with 100x100 mm^2 or 350x350 mm2 stroke from LPKF11, resolution better 200 nm.
- Nanorobotics modules, Microgripper, Nanofinger, Waferprober.
- Two rotation stages to change the orientation of the source and target area.

- Adhesive dispenser, Nanomanipulators, Scanning Probe Microscopes or e.g. a MicroProf from FRT.
- Video cameras and vision software for pattern recognition with a resolution of up to 400 nm used for orientation detection and quality control. Up to 6 cameras can be included.
- Process control with graphical user interface.

OPERATION MODES AND VISION SYSTEM

- The scale-up process is not only made possible by a modular hardware concept. The way from manipulation by hand up to automatic production with vision control is also possible.
- The manual mode allows to control all axes of the system with up to 2 (force feedback-) joysticks. Each joystick can move 4 different axes without changing the configuration.
- The second level of process control allows to generate sequences over menus and teaching with the manual control programme, a programming language is not necessary.

In theory microparts are delivered well oriented in carrier structures. But the reality delivers more something like micro chicken food. A flexible assembly system should also be able to handle these objects. Sorting mechanisms to orient these small parts in a flowing process are not always useful, because the microparts stick at these alignment structures by adhesion. Therefore the third level of process control allows to generate a closed loop process including vision control. Up to 6 video microscopes can grab pictures, and a pattern recognition software determines the position of not oriented microparts. Therefore a few millimeter picture size are necessary to keep an overview. At a field of view of 4x3 mm^2 the pattern recognition operates with a repeatability of 400 nm.

The Server PC with the vision software includes a command parser that allows a remote control over Ethernet commands from the Client PC. The client PC can switch between different cameras and ask for the position of recognized objects. The transformation from the coordinate system of the video camera to the world coordinate system is included and invisible for the user. Other functions of the system are e.g. the automated generation of a three-dimensional adhesive beads.

For example, when conductive adhesive is dispensed this line can be used as electrical wire to support a micro-device with power or signals. Furthermore, arrays of operations can be executed for parallel production. Automatic intensity adjustments are included, e.g. to couple a glass fibre into a device.

PROCESS CONTROL

The Micro Production System is controlled by a network of computers. A "Client-PC" is responsible for the global process as a master. A PC with a vision system as well as some embedded server PCs act as slaves and react only on commands from the master. The communication between these parallel operating devices is made by Ethernet. This structure has two main advantages:

- Flexible and easily configurable by adding or removing Ethernet-devices.
- Fast and reliable by parallel operation on different levels.

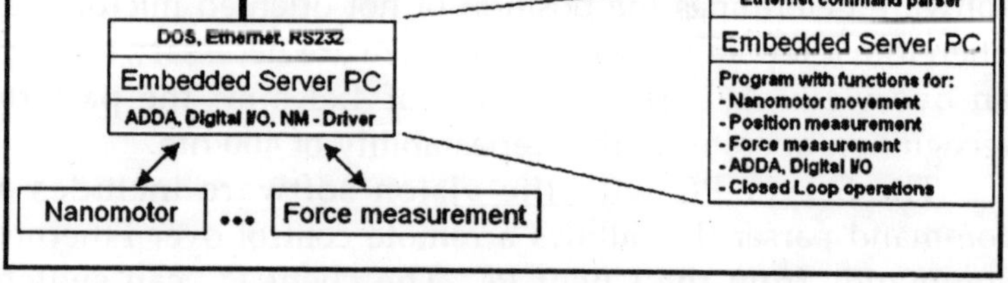

Fig. Scheme of the Ethernet based process control

The attempt to solve a set of tasks for micro production technology led to the development of a complete Micro Production System including Nanorobotics, that is up to 100 times more precise than classical production systems. The modular setup allows a scale-up for the production and an easy exchange of equipment, like at a CNC machine. Predefined software modules like vision controlled movement, programming of arrays or e.g. 4 axis synchronous movements allow to control complex procedures easily.

The Micro Production System can be used to develop new high resolution production processes, also as a service – including the development of adhesive joints.

Chapter 4

A New Medical Technology

Living systems can usually heal and cure their own injuries, unless those injuries are severe enough to prevent the living system from functioning. Too often, we suffer injuries that are indeed this severe. Molecular nanotechnology is feasible. As we master the ability to design molecular machines that can continue to function when the living system around them has failed, those molecular machines can restore the function of the living system. They can support and sustain the processes of the living system until that living system can once again function on its own.

Whether this is done by a temporary assist from respirocytes or by any of the myriad other techniques discussed in *Nanomedicine,* the underlying message is clear: life and health can be restored and sustained in the face of greater injury, greater damage, greater trauma, and greater dysfunction than has ever before been realized. This will usher in a new era of medicine — an era in which health and long life will be the usual state of affairs while sickness, debility and death will be the mercifully rare exceptions.

The future capabilities of nanomedicine give hope and inspiration to those of us who still have decades of life to look forward to, but some are not so fortunate. Many others who rightfully should live several decades more might find that chance cuts short their expected time. Heart attacks and cancer can strike us down even in the prime of our lives. They do not always wait their turn and politely arrive only when expected. How can today's dying patient take advantage of a future medical technology that is as yet only described in a handful of theoretical publications? How can we preserve the physical structure of our bodies well enough to permit that future medical technology to restore our health?

The extraordinary medical prospects ahead of us have renewed interest in a proposal made long ago: that the dying patient could be frozen, then stored at the temperature of liquid nitrogen for decades or even centuries until the necessary medical technology to restore health is developed. Called cryonics, this service is now available from several companies. Because final proof that this will work must wait until after we have developed a medical technology based on the foundation of a mature nanotechnology, the procedure is experimental.

We cannot prove today that medical technology will (or will not) be able to reverse freezing injury 100 years from now. But the patient dying today must choose whether to join the experimental group or the control group. The luxury of waiting for a definitive answer before choosing is simply not available. So the decision must be made today, on the basis of incomplete information. We already know what happens to the control group. The outcome for the experimental group has not yet been confirmed. But given the wonderful advances that we see coming, it seems likely that we should be able to reverse freezing injury — especially when that injury is minimized by the rapid introduction through the vascular system of cryoprotectants and other chemicals to cushion the tissues against further injury.

The development of nanomedicine depends on us: what we do and how rapidly we do it. Research is not done by a faceless "them," nor is it something that happens spontaneously and without any human intervention. It is done by and supported by people. Unless we decide to support and pursue this research, it won't happen. How long it takes to develop depends on us. We are not idle bystanders watching the world go by. We are a part of it.

If we sit and wait for someone else to develop this technology, it will happen much more slowly. If we jump in and work to make it happen, it will happen sooner. And developing a life saving medical technology within our lifetimes seems like a very good idea certainly better than the alternative.

Sometime in the next 30 years, mankind should acquire unprecedented ability to grasp, manipulate and modify individual molecules, and this ability may have profound implications. The term "molecular nanotechnology" has been coined to mean the technology of highly versatile and inexpensive molecular fabrication, molecular manipulation, and molecular-level manufacturing. It refers to the nanometer size range, the scale of atoms and molecules, but could be used to create macroscopic structures of precisely defined composition.

The key concepts of molecular nanotechnology are that molecules can be machines and that molecular engineers can work with molecules to build desired equipment just as effectively as today's engineers now work with bulk materials. The difference between nanotechnologists and biotechnologists is that the former do not restrict themselves to the biological limitations of the latter, and they are much more ambitious about the kinds of accomplishments that they want to achieve.

Organic chemists have long been able to synthesize complex molecular structures, including drugs, by devising ingenious reagents and by contriving reaction conditions in such a way as to minimize undesired side reactions and maximize yield. The results obtained depend on the statistics of uncontrolled molecular collisions in solution involving all possible molecular degrees of freedom. Statistically improbable reactions may be slow. More important, solution-based reactions generally lead to unwanted and potentially toxic byproducts.

This imprecision of present day chemical synthesis and manufacturing processes is one reason that drug standards are needed. In contrast, molecular nanotechnology would directly and rapidly produce the desired chemical and only the desired chemical, giving yields of 100 per cent. This would be achieved by precisely positioning and bringing together the individual molecules involved in the reaction in such a way as to catalyze the reaction desired and only the reaction desired.

MOLECULAR NANOTECHNOLOGY

Molecular nanotechnology has many precedents. Enzymes are natural molecular machines that adsorb individual reactant molecules from the surrounding solution and, as a result of precisely orienting them with respect to each other in a protected "nanoenvironment," catalyze reactions in a highly specific manner at very high speeds and under mild reaction conditions.

This simple process in biological systems ultimately allows synthesis of structures as diverse as carbon dioxide and hair. In fact, living organisms are naturally-existing, fabulously complex systems of molecular nanotechnology. If nature can produce the biochemical capabilities of living cells by accident, molecular engineers should be able to accomplish comparable, but broader capabilities by design, guided in part by the examples provided by living systems.

Many industries already make use of enzymes to catalyze desired reactions one molecule at a time. Genetic engineers are producing pharmaceuticals by using naturally occurring enzymes to edit DNA, and the soft drink industry uses enzymes that have been modified to allow them to produce sugar at high rates near 100 degrees C without denaturation. The real promise for the future, however, lies in the development of fully artificial enzymes. Enzymes have already been designed, synthesized, and found to function as designed.

Designed enzymes that are found not to function as intended can be modified as many times as necessary until they function as desired. Thus, whether from first principles or from enlightened trial and error, industrially useful artificial enzymes should be forthcoming. More than 10^{100} average-sized synthetic enzymes are possible with the use of nature's 20 amino acids, whereas probably less than 10^{14} enzymes are presently responsible for maintaining the entire biosphere.

But there is nothing to restrict artificial enzymes to only 20 amino acids. There are many useful kinds of chemistry that are not easily promoted by natural amino acids. One particularly versatile method for transcending the biological limits of proteinaceous catalysts has already been demonstrated.

It is based on the fact that the genetic code specifies amino acids by the sequence of any one of four nucleic acid bases taken three at a time, with each three-letter base sequence known as a codon. There are 64 codons in all, but nature uses them to specify only 20 amino acids rather than 64.

Recently, it has become possible to create an artificial transfer RNA that can add an unnatural amino acid to a growing polypeptide chain in response to one of the "unused" codons that exist naturally. This could be expanded to an unlimited number of unnatural amino acids, and these unnatural amino acids could contain totally nonbiological catalytic groups or even pre-made machine parts, such as structural support struts, molecular bearings, or the like. In a fully artificial system, the 20 natural amino acids might even be entirely dispensable.

Furthermore, it is now possible to insert artificial bases into DNA and RNA, drastically augmenting the prospects for designing catalytically active RNA as well as proteins with unnatural functional groups. The potential for programming the creation of unprecedented chemical catalysts and other molecular tools useful for molecular engineering is thus virtually open-ended.

Yet a ribosomal pathway to molecular nanotechnology is far from the only viable approach. Bulk techniques are now creating useful building blocks for molecular nanotechnology, such as carbon closed-end tubes with walls that are one atom thick. Cryptands or cavitands are being created that pack the catalytic punch of enzymes into nonprotein-aceous structures far smaller than natural enzymes.

At the same time, the ability of the scanning tunneling microscope not only to image, but also to manipulate individual atoms and molecules is being combined with the biological specificity of antibodies in an attempt to make all-purpose molecular synthesizers similar in concept to industrial robots that now make automobiles. These present-day efforts are being supplemented by computational chemistry simulations of molecular bearings, molecular planetary gears, and molecular robot arms that would direct molecular factories.

A New Era Of Medicine

An important premise of molecular nanotechnology is that the products of this technology should ultimately be readily affordable. This premise is based on the notion that an all-purpose molecular assembler should be able to make copies of itself. Thus, once the first programmable molecular assembler is made, it will not be long before there are as many such assemblers as the market needs.

Designs for computers with molecular data storage and processing elements predict data storage capacities of 1,000 megabytes per cubic micron and data-processing capabilities of 10^{10} operations per second for similar volumes. Translated into a sugar-cube-sized personal computer, this molecular computer would store 10^{20} bytes of information and process information at a speed of 10^{20} operations per second. This is 200 billion times the information storage capacity of today's top level personal computer and well over 100 million times the processing speed of today's fastest supercomputers.

Although it is difficult to project how many of the open-ended possibilities suggested by molecular nanotechnology will be realized by the year 2020, several innovations seem worth considering.

Quality control in the pharmaceutical industry today is necessary due to the imprecision of the manufacturing, purification, and packaging processes and the inability to monitor product quality on a continuous basis.

When drugs are made by programmable molecular fabricators and when molecular sensing devices are available, continuous sensing of the near-flawless production process should be possible, both for internal use and for reporting purposes. Problems could be corrected instantly by replacing defective chemical synthesizers with backup copies and discarding the few molecules of mis-synthesized material before any has a chance to leave the factory.

Thus, the products reaching the public should conform to standards with virtually exact fidelity. It is also possible that multipurpose drug synthesizers could be on hand in most clinical chemistry and toxicology laboratories and would make on-site *de novo* synthesis an alternative to obtaining pre-synthesized standards from bodies such as the USP. The barriers to this route could be more political than technical, but the availability of rapid local synthesis could make a life-or-death difference in some cases, such as those involving acute poisoning, for example.

Alternatively, and more probably, instruments capable of identifying and quantitating drugs without calibration standards should be feasible.

"Smart" Pharmaceuticals

Today's drug is essentially a single molecule with an often sophisticated but always limited repertoire. Tomorrow's "smart pharmaceuticals" could be essentially programmable machines with a range of "sensory," "decision-making," and "effector" capabilities.

They might avoid side effects and allergic reactions by coming in generic, biocompatible housings; becoming active only upon reaching their ultimate destinations; and attaining almost complete specificity of action. They might check for overdosage before becoming active, thus preventing accidental or intentional poisoning.

They might have not one chemical action but several, processing targeted invading organisms or malignant cells through a series of chemical reactions that guarantee the death of the target. They might work in concert with three or four "sister" agents that together produce versatility unattainable by one agent alone.

Despite the vast increase in complexity over present-day drugs, such agents can be expected to be totally "pure" and predictable in their behaviour.

Safety and efficacy may be inherent in the designs of such

"drugs," in which case regulatory issues could be simplified rather than complicated.

Drug and Health-Related Information

When every intelligent person can have essentially unlimited data storage and data-processing capability at his or her fingertips, drug information will change dramatically.

Marked advances in diagnostic agents could have far-reaching consequences. The year 2020 will occur approximately 15 years after the complete human genome has been obtained.

By 2020, great advances in understanding biochemical individuality, which is so important for side effects and proper dose adjustment, will have been made. This may make it possible for the physician to read critical aspects of a patient's phenotype from a noninvasively obtained cell sample and to inject "smart" diagnostic agents that can be recovered in a drop of saliva some time later and read out to reveal signs of previously undetected, impending disease processes. The opportunities for precise tailoring of individual treatment and for preventive medicine, with all the cost savings implied by both, would be revolutionary.

"Drug information" could come to include the complete matrix of appropriate pharmacologic responses to the newly-available individuality data.

Eventually, this will become so complex and extensive that the physician will be utterly dependent on assistance provided by computer "expert systems" in making decisions. In the long run, particularly if ingestible diagnostic agents and home readout systems become available, the need for physician participation in pharmacological therapy may vanish altogether for those patients with the right equipment available.

A major incentive for this decentralization of medical care will be ethical concerns over patient privacy and the use of genotypic information for unauthorized purposes: these are issues of "drug information" that continue to redefine the term. Relevant compendia should be available on line to any interested party, particularly those whose software allows them to interface the data with a medical "expert system."

Information about "drugs" per se will also change dramatically. Drug documentation will come to resemble today's protein structural databases because of the potentially great complexity of "pharmaceutical agents" that differ from medical devices only in that they are injectable, and/or invisible to the naked eye. Indeed, the line between pharmaceuticals and medical devices will become fuzzier and fuzzier. On the other hand, the need to document contra- indications, side effects, and complications of use should be considerably reduced.

The world of health care technology will be substantially different in 2020 from that of today. The Pharmacopeial Convention will likely be a much different organization on the occasion of its bicentennial anniversary.

The need for pharmaceutical scrutiny by medical practitioners could be even more intense, however, as all aspects of pharmaceutical medicine and the pharmaceutical industry itself continue to change dramatically and at great speed. It can be hoped that molecular nanotechnology, in addition to helping to create these dramatic changes, will also be one of the technologies that help the practitioners of 2020 stay abreast of and manage these developments.

CURRENT TECHNOLOGY

Currently the area of nanomedicine showing the greatest potential is the use of tiny polymeric drug carriers. Such materials can now be made in the form of spheres with diameters ranging from about 50 billionths of a meter. This can facilitate the transport of drugs to some of the smallest capillaries in the body.

These nanospheres are designed to travel to specific sites within the body, release their payload of drug molecules and then degrade.

Their degradation by-products are non-toxic and will ultimately be excreted from the body. Developments in polymer science have had a great influence on drug delivery.

It is now possible to synthesise a wide variety of biocompatible biodegradable polymers that will release entrained drugs at a rate determined by the chemistry and physical form of the polymer. Drug delivery research is now at a very important stage in its development.

In recent years there has been revolutionary progress made in the field of healthcare. The knowledge now exists to cure many afflictions that were once thought incurable through the use of novel therapeutics.

Many of these therapeutic methods cannot currently be used or are not being used to their full potential because they cannot be delivered into the body with the required control. An example of this is chemotherapy for the treatment of cancer. This involves saturating the body with toxic drugs, which can often result in harmful side effects such as reduced immune response to infection.

By directing the drug molecules specifically to the site of a tumour a reduced quantity of drug would be required thus reducing the toxic effect on the body.

Advances in biotechnology now make it possible to control disease at the genetic level. These developments have lead to the creation of several new fields in the area of biopharmaceutical science.

These new therapies are based on the interaction of nucleic acid drugs with the genetic material of specific cells within the body. The Achilles' heel of these new biopharmaceuticals is that it is difficult to direct them to the specific site required for action. There is also another disadvantage to these therapies in that they are easily broken down by conditions within the body. Incorporation of such drugs into delivery devices may be the way forward.

So how are such nanoparticulate devices produced? The answer is a complicated one that involves the use of molecules that self-assemble themselves during a chemical process. This procedure, which is known as 'emulsion polymerisation'. This involves the use of self-assembled spheres as a molecular 'mould' into which reactive monomer diffuses and polymerises to form solid spheres.

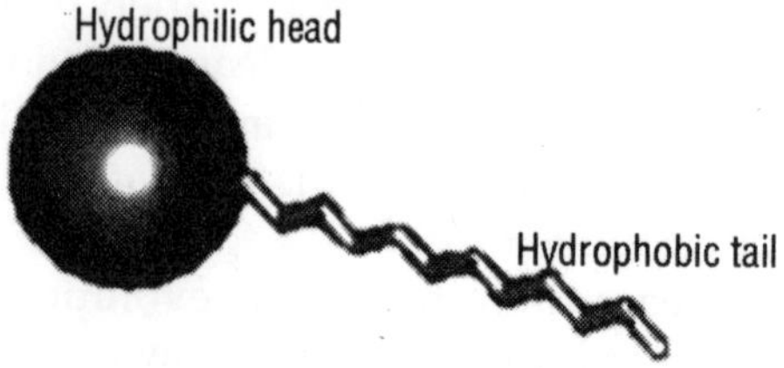

Fig. Schematic Representation of a Surfactant Molecule

Molecules known as surfactants perform this bizarre self-assembly by virtue of their unusual structure. The process has been likened to putting transistors, resistors and diodes into a bag, shaking it around and pulling out a fully functional radio. The mechanism that causes surfactant molecules to self-assemble is a result of their amphiphillic nature.

This means that the molecules are composed of two or more parts with each part being soluble in a different medium. Generally, a surfactant may have a hydrophilic head and a hydrophobic tail.

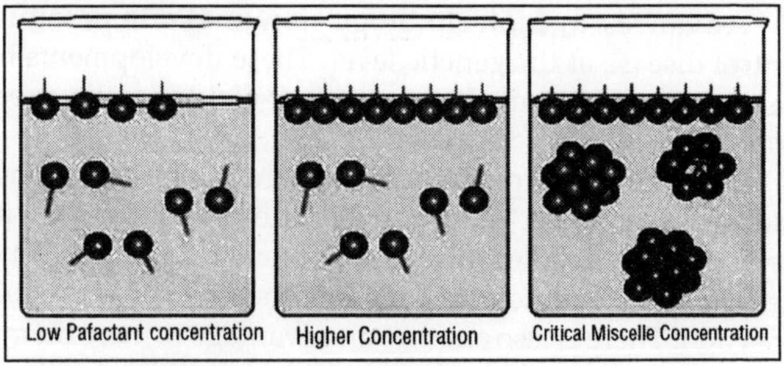

Fig. The Self Assembly Mechanism of Micelle Formation with Increasing Concentration of Surfactant

Once the concentration of these molecules in an aqueous solution reaches a critical value they will arrange themselves so as to minimise the interaction of the hydrophobic tail with the water. In order to accomplish this the hydrophobic tails align themselves along side each other that, in some cases lead to the formation of a spherical structure called a micelle. The size of these micelles is dependent on the length of the surfactant molecule and range from 10 - 100 billionths of a meter in diameter. Once there are millions of micelles spinning around in the aqueous medium a reactive monomer is introduced which is also hydrophobic.

Monomer molecules want to minimise their interaction with the water and for this reason they travel to the interior of the micelles where the are stable. The micelles will then swell to accommodate the monomer.

The final part of the preparation of the nanoparticles involves the addition of initiator molecules that trigger a chain reaction within the micelles core leading to a polymerisation. Once the core of the micelles have been converted into a solid polymer sphere of approximately 200 - 1000 billionths of a meter in diameter the spheres may be removed from the emulsion and are ready for use.

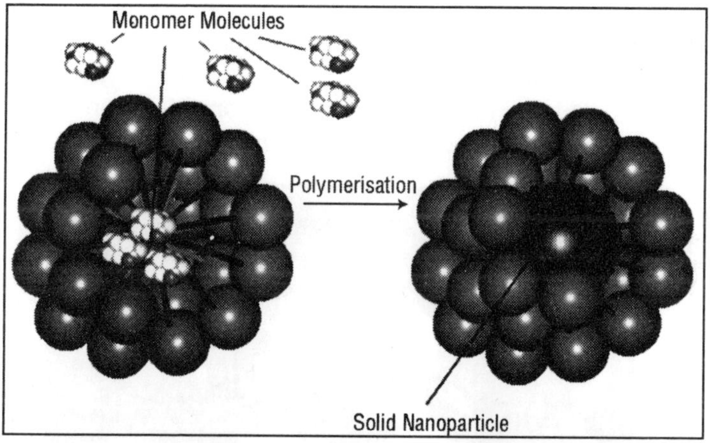

Fig. Once Monomer Molecules Diffuses into the Micelle they are Polymerised to form a Solid Nanosphere

Once produced these tiny particles appear as a fine powder that can have drug molecules either attached to their surface or absorbed into their core. At this stage the nanoparticles are solid spheres composed of polymer chains of a fixed length. This chain length is described by the polymer's molecular weight. For a specific polymer, an example being poly(n-butyl cyanoacrylate), the lower the molecular weight the faster it will degrade within the body and thus the faster the entrained drug is released. The specific site within the body to which the nanoparticles will travel is controlled by the surface chemistry of the nanoparticle. Molecules can be tethered to the particles' surface, which will cause selective uptake, by various organs within the body.

FUTURE RESEARCH

Advances in analytical techniques have improved our ability to produce and image nanostructures. Scanning Electron Microscopy (SEM) uses an electron beam to image nanoparticles. They cannot be seen using standard optical microscopy because their diameter is smaller than the wavelength of visible light.

SEM however has its disadvantages in that the powerful electron beam used can often vaporise the particles. A new technique is now becoming available that generates an image by passing a stylus back and forth over the surface of the particles. This stylus is much like the one used on a record player but is a few thousand times smaller.

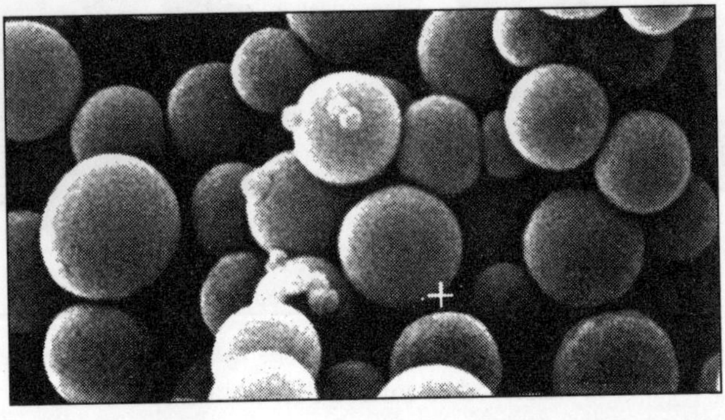

Fig. Scanning Electron Microscope image of nanoparticles

The technique is called Atomic Force Microscopy (AFM) and its magnifying capability is such that it has been compared with the ability to count the grains of sand on a beach on earth, from the moon! At the moment there is a great deal of research being carried out in the area of nanotechnology.

Materials that change their physical characteristics

following exposure to stimuli such as light, heat and changes in pH are being developed. It has been proposed that such materials will lead to the first nano-scale robot arms and similar devices. By combining technologies it may be possible to create 'smart' nanoparticles that will be the next step towards true cellular level medicine.

NANOMEDICINE: THE BIGGER PICTURE

Future techniques in medical diagnosis and treatment have often been the subject of science fiction literature and cinema. The approaches are surprisingly consistent. In Star Trek, for example, the diagnosis of illness and its treatment is generally non-invasive and mostly painless.

In 'The Six Million Dollar Man', an individual who has undergone serious accidental damage is given replacement organs and tissues that function as well, or even better than the originals. What was once the stuff of science fiction is now closer to becoming a reality.

Nature operates at the nanoscale, and today we are acquiring an increasingly profound understanding of natural processes at this scale, enabled by a new generation of scientific instruments. From this knowledge, we are able to design devices that can either directly interact with, or influence, the behaviour of living cells. Nanotechnology has a trump card to play when applied to medicine. At the nanometre scale, materials often exhibit surprisingly different physical, chemical, and biological properties when compared with the same material in bulk form.

The properties of nanoparticles, such as increased chemical activity and the ability to cross tissue barriers, are leading to new drug targeting and delivery techniques. In the future, it is within the realms of possibility that a nanoparticle may be designed to search for, find and destroy a single diseased cell, taking us even closer to realizing the ultimate goal of disease prevention.

Nanotechnology is also making possible other techniques, such as the stimulation of the body's own mechanisms to successfully repair diseased or damaged tissues, replacing the need for transplants and artificial organs. In the forseeable future, allied to advances in information and communication technologies, nanotechnology as applied to medicine, will lead to advances in remote monitoring and care, where a patient may be treated at home - a less expensive option, and one that is more conducive to a successful medical outcome than treatment in a surgery or hospital.

Continued research into disease processes at the molecular level is essential for the development of nanomedicine, and involves teams of scientists from across 'conventional' disciplines, such as physics, chemistry, surgery and mathematics, as well as those from the 'new' fields of genomics, proteomics, metabolomics, pharmacokinetic modelling and even microscope design.

Challenges exist in training and managing these multidisciplinary teams and partnerships, and importantly, in finding new solutions to the intellectual property issues that presently hinder the speedy commercialization of new knowledge.

Research challenges also relate to understanding and modelling the toxicity of engineered nanoparticles, and requires that toxicologists work alongside medical scientists wherever nanoparticles are involved in drug targeting and contrast agent development. There are technological challenges, too, in the areas of molecular manufacturing, quality assurance and the eventual programmability of nanodevices.

Nanomedicine will also bring its own set of new legal and ethical challenges to be resolved. Cost benefit analyses are also critical. Where money can be best invested for the greatest social and economic benefit needs to be carefully decided. Should investment be made in prevention or treatment; in diseases of the poor or the rich; in long term hospitalization or costly drug development and deployment?

Any strategy for nanomedicine must also be influenced by the facts that resources are finite and demands are great. Within these constraints the health needs of an ageing population need to be managed, cures found for major lifestyle 'killer' diseases, such as cancer, and for the diseases of the less-developed world, such as HIV/ aids, malaria and tuberculosis. In conclusion, nanoscience and nanotechnology are leading to extraordinary new breakthroughs in medicine that were once the stuff of dreams. EuroNanoForum aims to provide a small but representative snapshot of these exciting opportunities.

TISSUE ENGINEERING AND NANOSCAFFOLDS

Science is moving from transplanted organs to implanting of substitute or artificial organs to stimulating the body to do its own repairs. Tissue regeneration is about enabling the body to ultimately regenerate its own diseased or failed organs. The market worldwide for tissue engineered products already worth $18bn, and many hundreds of thousands of people are awaiting transplants worldwide. This has led to an illegal market in donor organs from the poor to the rich, so the development of effective regeneration techniques offers a multitude of benefits.

Tissue Regeneration Work

Instead of being surgically repaired, transplanted, or even fixed using prosthetics, tissue or organ failure could be solved by implanting natural tissue and organ mimics which can be fully functional from the start, or grow into the required functionality. Nanotechnology helps to recruit the body's natural healthy cells to promote regeneration of tissue on or around a damaged area.

It can also act as a 'scaffold' to provide a framework for developing tissues to latch onto and penetrate. This application of nanotechnology will result in reconstructed tissue and wound treatments that are superior, longer lasting and more acceptable to everyone involved, most notably the patient.

The beauty of the process is that it is arguably the most natural way of healing, as it is the body's own healthy cells which are regenerating - as they were meant to do - albeit with a small push to get started. Tissue regeneration would be almost unthinkable without development in nanoscience and nanotechnology. It underpins the design of scaffolds on which the cells are grown, which are composed of special nanocomposite materials hat contains cell growth stimulants, and have nanoscale, cell-friendly surface topographies.

There is even the potential for whole organs to be grown to replace those that have failed through disease or old age.Tissue regeneration offers a revolution in healthcare, with huge benefits for doctors who will be using the technology, to the government who will save money, and most importantly to the patients who are the ultimate beneficiaries.

PHARMACEUTICAL DEVELOPMENT

A long standing issue for pharmaceutical companies is to deliver the right amount of drug to the site of disease. The inability to achieve this has means that drugs need to be administered in excessively high doses, increasing the likelihood that patients may suffer from toxic side effects.

Nanotechnology enables drugs to be targeted more effectively to the site of the disease, and activated 'on arrival'. Drug can be prepared in such a way that they can cross epithelial barriers, and differentiate between healthy cells and diseased ones. Very small drug particles (nanoparticles) are coated with special molecules and travel round the body until they reach the disease site.

Once there, the molecules 'decorating' the drug particles recognise the disease site, bind to it, and the drug is then activated. This targeting means the drug will work efficiently and more effectively, and side-effects dramatically decrease. For example, chemotherapy in cancer usually means healthy cells are also killed off during the attempt to kill the cancerous cells. Targeted drug delivery circumvents this, and offers the potential for a quicker recovery.

Nanotechnology has other benefits for therapy. In nanoparticulate form, drugs are less prone to degradation, meaning they will have a longer shelf-life. Nanoparticluate drugs can be coated to delay their activation, resulting in the drug remaining in the body longer, reducing the frequency of dosage required. Some drugs may be linked to particles which fluoresce when they reach diseased cells. This is a special technique known as find, fight and follow, and is leading to early detection and treatment.

The area of disease can be identified by imaging the fluoresecent particles, which then deliver their payload of drugs, and the effectiveness of the drugs can be then monitored. In essence, the application of nanotechnology to drug targeting and delivery offers more effective treatments, lower toxicity and side-effects and reduced cost from reduced dosage. Drug delivery and targeting are areas which are already offering real patient benefits.

CELL STRUCTURE AND FUNCTION

Humans are essentially a collection of cells! The better we understand how cells work, the more likely it is we can create effective cures, understand the cause of disease better and produce better drugs. Nanotechnology is enabling this understanding. This session is concerned with understanding of cell structure and function and how a cell manifests disease, and the new tools that are being developed to identify early signs of disease at the cellular level, monitor of healthy and diseased cells and the progress of therapy, and measure the effects of different drugs.

Once we know the signals that a cell gives when it is diseased we can look for that signal. Handin-hand with the growing understanding of cell function, is the technology for measuring and identifying changes that indicate disease. The development of 'lab-on-a-chip' is one such technology which has many applications in disease identification.

These include enabling cells to be studied independently of the body, the speedy screening patients for infection, and the analysis of the hereditary propensity to disease. A lab-on-a-chip is a miniature laboratory that uses techniques developed from the electronics industry, that offers speed, sensitivity, accuracy and portability, compared with more traditional techniques, which are slow, expensive resource intensive and may be inaccurate as samples deteriorate with time.

In the future, this technology is leading to patient-centred medicine, with patients conducting tests in at their own homes, and downloading the results via their computer to a medical monitoring centre.

Lab-on-a-chip technology also has important applications in being able to quickly screen a large population for disease or potentially killer infections, such as avian flu. Lab-on-a-chip techniques and a knowledge of the human genome is leading to pre-symptomatic disease treatment through identify the individual genetic propensities say to breast or colon cancer or heart disease, so appropriate preventative action can be taken.

This technology is also important for the developing world specifically in some African countries where disease is at epidemic rates. In areas where very little money, or even healthcare is available, the provision of cheap, fast-acting LOCs could offer early identification of disease, and enable more appropriate effective and cheaper treatment options. The cost effectiveness of what is basically a tiny microchip is another selling point; it can be mass produced incredibly cheaply.

THE PROMISE OF NANOMEDICINE

Nanomedicine offers a huge spectrum of benefits for the better treatment of disease – from early diagnosis to better therapies to eventually preventative medicine; from stimulating the body to regenerate its own diseased tissues to the targeting and activating drugs at the site of diseae. It also offers the possibility of developing better drugs faster, that have fewer side effects and are customised to the patient's own genome.

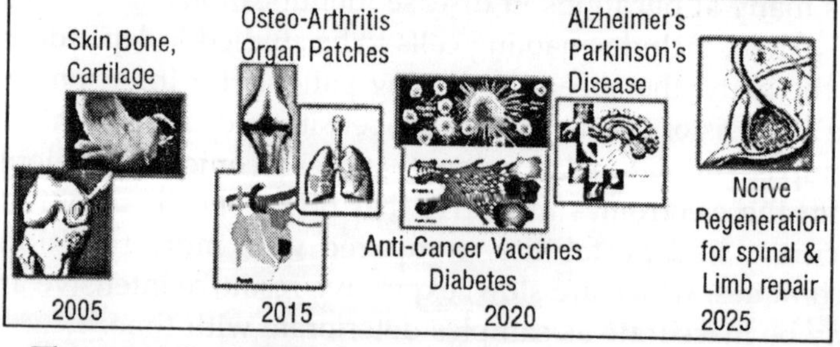

Three particular areas where nanotechnology will have a major impact in relation to cancer; in relation to reducing animal testing, and in the commercial impact of new nano-based therapies:

Cancer treatment is the past has suffered from a lack of understanding the mechanisms of the disease at the cellular level, late diagnosis, and treatments based on attacking the whole body rather than the disease site. Nanotechnology is offering a totally new approach; from a better understanding of the root causes of disease, to quicker diagnosis and science-base treatments. Today, to quote Dr Mauro Ferrari 'we are seeing the tip of the iceberg' in terms of the potential nanotechnology offers for cancer treatment.

Another area where we are using a hammer to crack a nut is in the area of animal experiments for the validation of drugs. This has obviously produced some benefits in the past, but these are relatively small in relation to the large body of research (and cost) as many apparently good results could not be adequately replicated in humans, or produced unexpected side effects. With a better knowledge of individual responses to chemicals, nanotechnology is leading to a refinement of techniques for drug discovery that are accurate and meaningful for the individual patient, based on the use of living cells.

Nanotechnology is leading to better treatments based on our knowledge of proteomics and genomics, this is leading to a winwin situation for drug companies and patients as well as huge economic benefits for those countries where the State provides healthcare. Drugs are becoming more efficient as they are increasingly designed to treat specific diseases, patients are cured more rapidly, less care is needed to counteract reactions.

APPLICATIONS OF NANOTECHNOLOGY IN MEDICINE

Nanotechnology is leading to faster diagnosis and more effective treatment, the early signs of disease, and its exact location. One such application uses quantum dots to locate the disease site. Quantum dots are semiconductor nanoparticles which can glow very brightly under certain criteria. The quantum dots can be attached to chemicals which link to diseased cells. When the link is made, the quantum dots glow. This allows the sites of disease to be accurately pinpointed, and also malignant and non-malignant tumours can be differentiated. Quantum dots may help in identifying cells which may develop problems in the future.

By recognising certain indicators that cannot be detected using conventional technologies, quantum dots can call attention to potential disease sites, leading to the ideal of disease prevention. Improved drug targeting and delivery is another highly promising area of nanomedicine, which is actually happening now.

Coated drugs in nanoparticulate form can be delivered directly to the diseased cells. This means smaller or less frequent doses are required, and toxicity is reduced as the targeted drug works more effectively. Therefore nanomedicine is leading to the reduction of undesirable and costly side-effects of drugs, as the customisation and targeting of treatment would allow for more control over where the drug goes and what it affects in the body.

Nanotechnology is set to revolutionise healthcare through moving us towards preventative, or pre-emptive medicine, with many health and cost benefits, both to the State and the patient. Impacts are especially envisioned for the ageing and other hitherto disadvantaged populations.

CONVERGING TECHNOLOGIES HEALTHCARE

Limits and barriers on medical research are being rapidly broken down by the convergence of many disciplines, such as nanotechnology, molecular biology with physics, information technology and even, at last, the humanities! Together, this is resulting in rapid advances in new techniques, processes and tools which are completely changing the way we have approached healthcare in the past.

It is not only researchers from different disciplines that are working together. Networks like Nano2Life are bringing private firms, scientists and research institutes together to solve the problems of treating disease. Strong partnerships are being formed that are providing a range of skills across the board, generating disruptive technologies and completely innovative solutions to old problems. These could include non-invasive diagnostics, individual-specific drugs, identifying disease at the cellular level, faster validation of new drugs, and stimulating the patient's own body to regenerate diseased or failed organs.

This session will examine convergence and its benefits on a European level. It is case studybased, describing networks and their component organizations, and the benefits that have accrued to industry from EU-supported partnerships and initiatives that connect complementary skills from a wide range of disciplines.

NANOTECHNOLOGY FOR DEGENERATIVE DISEASES

Nanotechnology is expected to play an increasing role in the battle against congenital and degenerative diseases. As we live longer, many more diseases become manifest over a larger proportion of the population, such as Alzheimer's and Parkinson's. We are also experiencing an epidemic of lifestyle diseases, from diabetes to cancer to heart and vascular disease.

Firstly, early diagnosis is critical to many of these diseases, which is enabled by the miniaturization, user-friendliness, cheapness and wide availability of diagnostic techniques to such an extent that patients can monitor their own symptoms. Example include self-testing for diabetes and cholesterol levels. The spectrum of possibilities will expand dramatically; test results can be transmitted to a central health bureau, and therapies prescribed remotely. Working at the nanoscale has produced new hope for disease of the brain, for example.

If the therapy is drug-based, drugs in nanoparticulate form can now cross the blood-brain barrier. If the therapy requires regeneration of diseased tissue, new research is enabling the growth of the patient's own cells on nano-inspired scaffolds that can be then implanted into the brain.

Other areas where Nanotechnology offers hope is in the delivery of drugs in a way that emulated the body's own systems. For example, in diabetes, insulin is required by the body at irregular intervals. Using techniques derived from the electronics industry, this need can be monitored using a small, silicon chip-based implant, and insulin delivered only as required. Electronics combined with nanotechnology is also the basis of better retinal and cochlear implants.

Today, new retinal implants have been developed that enable people previously classified as blind to be reclassified as sighted. Present cochlear implants are crude, difficult to insert, awkward to wear and give poor performance, Future implants will use nanocomposite materials, be small and easy to insert and be cheap to produce. This also has very promising implications for the poor.

Nanoimaging and Diagnos

Medical imaging has advanced from a marginal role in healthcare to becoming an essential tool of diagnostics over the last 25 years. Molecular imaging and image-guided therapy is now a basic tool for monitoring disease and in developing almost all the applications of in-vivo nanomedicine. Originally, imaging techniques could only detect changes in the appearance of tissues when the symptoms were relatively advanced. Later, contrast agents were introduced to more easily identify and map the locus of disease.

Today, through the application of nanotechnology, both imaging tools and marker/ contrast agents are being dramatically refined towards the end goals of detecting disease as early as possible, eventually at the level of a single cell, and monitoring the effectiveness of therapy. The convergence of nanotechnology and medical imaging opens the doors to a revolution in molecular imaging, leading eventually to the detection of a single molecule or a single cell in a complex biological environment.

Engaging The Community

Nanomedicine offer many benefits for healthcare in the future. It is a complex subject, and additionally suffers from many misconceptions about the potential of Nanotechnology for medicine, arising out of a confusion between science fiction and fact, even amongst highly educated people. Considering governments and institutions the world over are spending many billions on nanotechnology research and development, a small amount of that funding would be beneficially channelled to wider public awareness and engagement.

As with any new technology, there are risks as well as benefits. Too frequently the downsides of an innovation are publicised, out of proportion to the benefits. A survey carried out last year by researchers at the University of North Carolina established that the more people knew about nanotechnology, the more they thought the benefits would greatly outweigh the risks. The opposite was also true.

This is a key point which indicates a wellinformed public is likely to embrace nanotechnology and, whilst remaining wary of some of the risks, if they see it as bringing a major benefit to their lives. To inform the public, so they can make reasoned and reasonable decisions, and what needs to be put in place to provide reassurance that potential risks are receiving the right level of attention and response.

DISEASES OF THE DEVELOPING WORLD

According to a study by the Canadian Programme on Genomics and Global Health (CPGGH) at the University of Toronto Joint Centre for Bioethics (JCB) - a leading international medical ethics think-tank - nanotechnology may provide the means to help developing countries across as spectrum of areas including improvement in water quality, reduction in environmental and early diagnosis and treatment and even prevention of killer diseases such as malaria, tuberculosis and HIV/AIDS.

Benefits of Nanotechnology

Diagnosis of disease: About a quarter of all Africans are infected with AIDS. It kills millions each year and 95 per cent of all new cases of AIDS occur in developing countries. Nano-based devices are leading to easy and fast analysis of a range of diseases that can be undertaken cheaply and easily.

Treatment of Disease.: Nanomedicine also has the potential to ensure longer shelf life for drugs, easier administration, and lower dosage requirements. New high throughput screening techniques are leading to cheaper drug discovery; and cell based assays should lead to faster introduction of these drugs into the population.

Sources of Disease. Nanotechnology-enabled systems have application for water purification through membrane filters, which would have the knock-on effect of reducing infection.

Other: Apart from benefiting medicine directly, nanotechnology is also offering access to cheap renewable energy using new polymer based solar power collectors that are lightweight, cheap and efficient.

Needs of the Ageing Population

Nanotechnology could finally mean that growing old does not necessarily mean loss of faculties and a rapid reduction in quality of life. Nanotechnology is supporting the rehabilitation of the infirm and elderly through the development of intelligent learning prosthetic devices. These include retinal and cochlear implants. Through the development of nanocomposites, nanoelectronics, nanosensors, nanotransducers and biomaterials, the next generation of implants will be more effective, body friendly and less traumatic to insert. They will reinvigorate the faculties of patients whose hearing or sight has been affected by accident, disease or age. Organ replacement is increasingly common across the generations.

Transplantation is fraught with dangers such as rejection and side effects from drugs. As discussed under Tissue Engineering, research is leading to the growth of new organs from a patient's own tissues. The ultimate objective is to produce innovative scaffold materials which canm be seede with the patient's own cells, which, when implanted will regenerate bone, cartilage and skin tissues in the most natural way possible. Medical textiles is an exciting field of innovation. 'Smart clothes'are being developed which will regulate temperature and even monitor the health of the wearer.

This remarkable new technology poses even more

questions regarding ethics – the ethics of who should benefit, if there is a choice to be made, where the line should be drawn, if any, at enhancing human performance, and whether new technology will lead to a further widening of the gap between rich and poor, advantaged and disadvantaged.

IMPLANTS AND DEVICES

Nanotechnology, new materials and nanoelectronics are the basis of novel retinal and cochlear implants. New retinal implants have been developed that enable people previously classified as blind (through disease, such as macular degeneration) to be reclassified as sighted. Present cochlear implants are crude, difficult to insert, awkward to wear and give poor performance, Future implants will use nanocomposite materials, be small and easy to insert and be cheap to produce. This also has very promising implications for the poor.

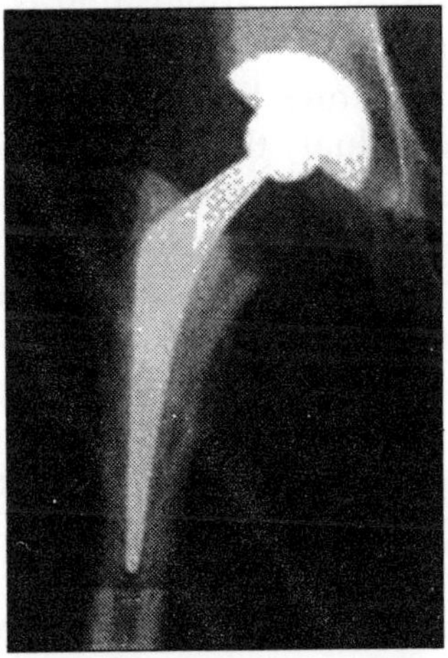

Over $30bn is spent on implants yearly (hip, knee, dental etc.). Nanotechnology can help bring about better implants. For example, Proteins at the body-implant interface play a decisive role in the acceptance or rejection of the implant.

Using nanotechnology we can produce 'bodyfriendly' coatings and also create topographies that cells like, to increase the likelihood of acceptance. These biocompatible coatings are even being used in devices like hearing aids,reducing the likelihood of irritation or skin reaction. Around 1,000,000 hip and knee-joint replacements are carried out in the EU each year. However, the lifespan of the implants is only around 10 years, shorter if the patient is particularly active or overweight. This poses both a quality of life problem, and a cost effectiveness issue. Revisionary surgery adds about €520m a year to the EU's medical costs. Nanocomposites are also being developed that form ceramic implants which are stronger and longer-lasting, with potential life-spans of more than 30 years.

NANOSENSORS AND DIAGNOSTICS

In nanodiagnostics, the ultimate goal is to identify disease at the earliest stage possible, ideally at the level of a single cell. When testing for disease, nanotechnology offers better sensitivity, specificity and reliability. Several steps from sample preparation to detection can be integrated into a single, tiny device. An important application is in cheap, reliable devices, based on nanotechnology, that can be used at the "point of care" in clinics by non-expert technicians, and eventually by individuals themselves. Advances in in-vivo diagnostics will rely mainly on imaging. In molecular imaging, the goal is to create sensitive, reliable agents that can detect, deliver and monitor therapy. This is the "find, fight and follow" concept, sometimes also known as *theranostics*.

The likely diseased tissue is firstly imaged, using target-specific contrast agents. These contrast agents can be combined with a drug; and the results of therapy can then be monitored, again using imaging techniques.

The advances in diagnostics have presented major opportunities for healthcare, including a plethora of new personalised treatments and early diagnosis for individuals suffering from a wide variety of illnesses such as cancer, diabetes, and heart disease. Nanotechnology has even made a tiny 'nano needle' possible which can provide important information on disease at the level of a single cell. Other types of nanosensors include cantilevers which are sensitive, simple and cheap and able to accurately detect tiny biochemical changes.

Chapter 5

Scanning Tunneling Microscopy

The scanning tunneling microscope (STM) is the ancestor of all scanning probe microscopes. It was invented in 1981 by Gerd Binnig and Heinrich Rohrer at IBM Zurich. Five years later they were awarded the Nobel prize in physics for its invention. The STM was the first instrument to generate real-space images of surfaces with atomic resolution.

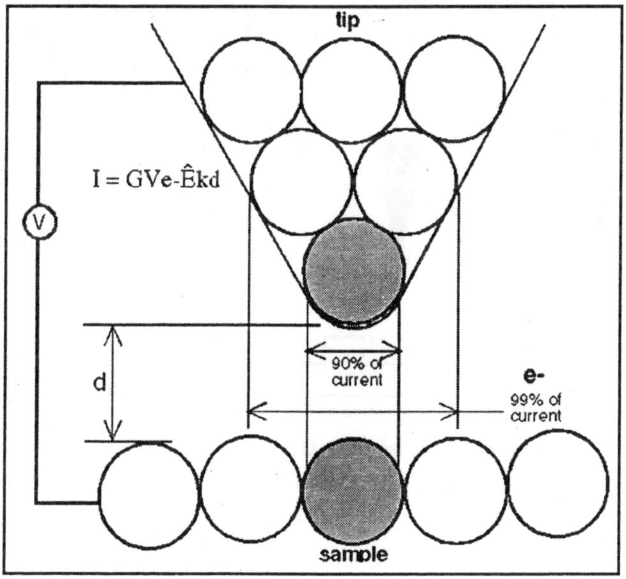

Fig. Schematic of Tip and Sample Interaction

STMs use a sharpened, conducting tip with a bias voltage applied between the tip and the sample. When the tip is brought within about 10Å of the sample, electrons from the sample begin to "tunnel" through the 10Å gap into the tip or vice versa, depending upon the sign of the bias voltage. The resulting tunneling current varies with tip-to-sample spacing, and it is the signal used to create an STM image. For tunneling to take place, both the sample and the tip must be conductors or semiconductors. Unlike AFMs, STMs cannot image insulating materials.

STMs can be designed to scan a sample in either of two modes: *constant-height* or *constant-current* mode.

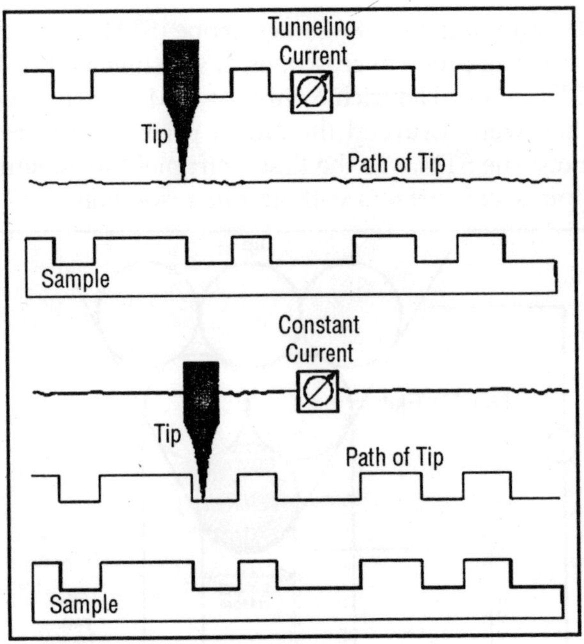

Fig. Comparision of Constant - Height and Constant-Current Mode for STM

In constant-height mode, the tip travels in a horizontal plane above the sample and the tunneling current varies depending on topography and the local surface electronic properties of the sample. The tunneling current measured at each location on the sample surface constitute the data set, the topographic image.

In constant-current mode, STMs use feedback to keep the

tunneling current constant by adjusting the height of the scanner at each measurement point. For example, when the system detects an increase in tunneling current, it adjusts the voltage applied to the piezoelectric scanner to increase the distance between the tip and the sample.

In constant-current mode, the motion of the scanner constitutes the data set. If the system keeps the tunneling current constant to within a few per cent, the tip-to-sample distance will be constant to within a few hundredths of an angstrom. Each mode has advantages and disadvantages. Constant-height mode is faster because the system doesn't have to move the scanner up and down, but it provides useful information only for relatively smooth surfaces. Constant-current mode can measure irregular surfaces with high precision, but the measurement takes more time.

As a first approximation, an image of the tunneling current maps the topography of the sample. More accurately, the tunneling current corresponds to the electronic density of states at the surface. STMs actually sense the number of filled or unfilled electron states near the Fermi surface, within an energy range determined by the bias voltage. Rather than measuring physical topography, it measures a surface of constant tunneling probability.

From a pessimist's viewpoint, the sensitivity of STMs to local electronic structure can cause trouble if we are interested in mapping topography. For example, if an area of the sample has oxidized, the tunneling current will drop precipitously when the tip encounters that area. In constant-current mode, the STM will instruct the tip to move closer to maintain the set tunneling current. The result may be that the tip digs a hole in the surface.

From an optimist's viewpoint, however, the sensitivity of STMs to electronic structure can be a tremendous advantage. Other techniques for obtaining information about the electronic properties of a sample detect and average the data originating from a relatively large area, a few microns to a few millimeters across. STMs can be used as surface analysis tools that probe the electronic properties of the sample surface with atomic resolution. Using STMs as spectroscopy tools is discussed later (Scanning Tunneling Spectroscopy).

ATOMIC FORCE MICROSCOPY

The atomic force microscope (AFM) probes the surface of a sample with a sharp tip, a couple of microns long and often less than 100Å in diameter. The tip is located at the free end of a cantilever that is 100 to 200µm long. Forces between the tip and the sample surface cause the cantilever to bend, or deflect. A detector measures the cantilever deflection as the tip is scanned over the sample, or the sample is scanned under the tip. The measured cantilever deflections allow a computer to generate a map of surface topography. AFMs can be used to study insulators and semiconductors as well as electrical conductors.

Several forces typically contribute to the deflection of an AFM cantilever. The force most commonly associated with atomic force microscopy is an interatomic force called the van der Waals force. The dependence of the van der Waals force upon the distance between the tip and the sample.

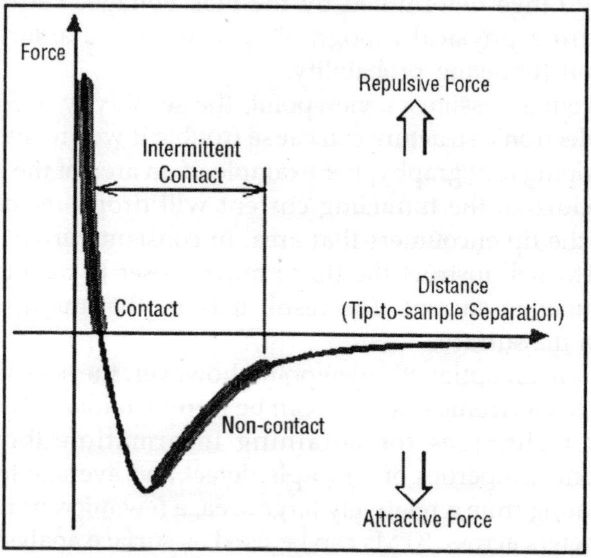

Fig. Interatomic Force vs. Distance Curve

Two distance regimes are labeled:

- The contact regime;
- The non-contact regime.

In the contact regime, the cantilever is held less than a few angstroms from the sample surface, and the interatomic force between the cantilever and the sample is repulsive. In the non-contact regime, the cantilever is held on the order of tens to hundreds of angstroms from the sample surface, and the interatomic force between the cantilever and sample is attractive (largely a result of the long-range van der Waals interactions).

CONTACT AFM

In contact AFM mode, also known as repulsive mode, an AFM tip makes soft "physical contact" with the sample. The tip is attached to the end of a cantilever with a low spring constant, lower than the effective spring constant holding the atoms of the sample together. As the scanner gently traces the tip across the sample (or the sample under the tip), the contact force causes the cantilever to bend to accommodate changes in topography. To examine this scenario in more detail, refer to the van der Waals curve.

At the right side of the curve the atoms are separated by a large distance. As the atoms are gradually brought together, they first weakly attract each other. This attraction increases until the atoms are so close together that their electron clouds begin to repel each other electrostatically. This electrostatic repulsion progressively weakens the attractive force as the interatomic separation continues to decrease. The force goes to zero when the distance between the atoms reaches a couple of angstroms, about the length of a chemical bond. When the total van der Waals force becomes positive (repulsive), the atoms are in contact.

The slope of the van der Waals curve is very steep in the repulsive or contact regime. As a result, the repulsive van der Waals force balances almost any force that attempts to push the atoms closer together. In AFM this means that when the cantilever pushes the tip against the sample, the cantilever bends rather than forcing the tip atoms closer to the sample atoms. Even if we design a very stiff cantilever to exert large forces on the sample, the interatomic separation between the tip and sample atoms is unlikely to decrease much.

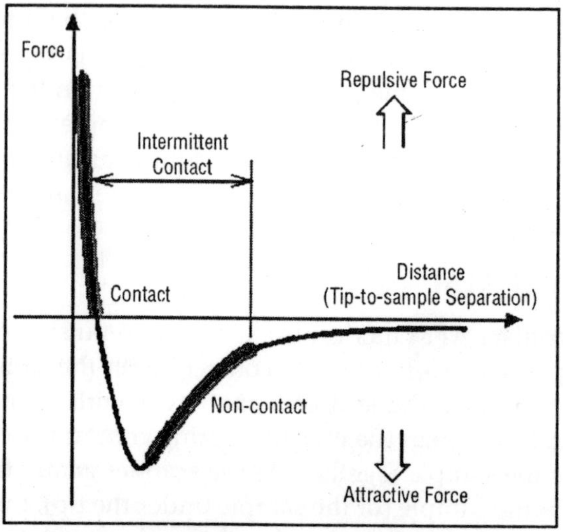

Fig. Interatomic Force vs. Distance Curve

Instead, the sample surface is likely to deform (Nanolithography). In addition to the repulsive van der Waals force described above, two other forces are generally present during contact AFM operation: a capillary force exerted by the thin water layer often present in an ambient environment, and the force exerted by the cantilever itself.

The capillary force arises when water wicks its way around the tip, applying a strong attractive force (about 10-8N) that holds the tip in contact with the surface. The magnitude of the capillary force depends upon the tip-to-sample separation. The force exerted by the cantilever is like the force of a compressed spring. The magnitude and sign (repulsive or attractive) of the cantilever force depends upon the deflection of the cantilever and upon its spring constant.

As long as the tip is in contact with the sample, the capillary force should be constant because the distance between the tip and the sample is virtually incompressible. It is also assumed that the water layer is reasonably homogeneous. The variable force in contact AFM is the force exerted by the cantilever.

The total force that the tip exerts on the sample is the sum of the capillary plus cantilever forces, and must be balanced by the repulsive van der Waals force for contact AFM. The magnitude of the total force exerted on the sample varies from 10-8 (with the cantilever pulling away from the sample almost as hard as the water is pulling down the tip), to the more typical operating range of 10-7 to 10-6N.

Most AFMs currently on the market detect the position of the cantilever with optical techniques. In the most common scheme, a laser beam bounces off the back of the cantilever onto a position-sensitive photodetector (PSPD). As the cantilever bends, the position of the laser beam on the detector shifts. The PSPD itself can measure displacements of light as small as 10Å. The ratio of the path length between the cantilever and the detector to the length of the cantilever itself produces a mechanical amplification. As a result, the system can detect sub-angstrom vertical movement of the cantilever tip.

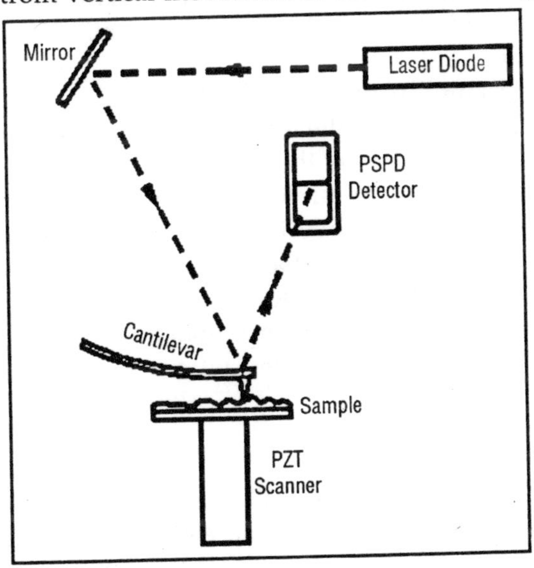

Fig. The Beam-Bounce Detection Scheme

Other methods of detecting cantilever deflection rely on optical interference, or even a scanning tunneling microscope tip to read the cantilever deflection.

One particularly elegant technique is to fabricate the cantilever from a piezoresistive material so that its deflection can be detected electrically. (In piezoresistive materials, strain from mechanical deformation causes a change in the material's resistivity.) For piezoresistive detection, a laser beam and a PSPD are not necessary.

Once the AFM has detected the cantilever deflection, it can generate the topographic data set by operating in one of two modes - *constant-height* or *constant-force* mode. In constant-height mode, the spatial variation of the cantilever deflection can be used directly to generate the topographic data set because the height of the scanner is fixed as it scans.

In constant-force mode, the deflection of the cantilever can be used as input to a feedback circuit that moves the scanner up and down in z, responding to the topography by keeping the cantilever deflection constant. In this case, the image is generated from the scanner's motion. With the cantilever deflection held constant, the total force applied to the sample is constant.

In constant-force mode, the speed of scanning is limited by the response time of the feedback circuit, but the total force exerted on the sample by the tip is well controlled. Constant-force mode is generally preferred for most applications.

Constant-height mode is often used for taking atomic-scale images of atomically flat surfaces, where the cantilever deflections and thus variations in applied force are small. Constant-height mode is also essential for recording real-time images of changing surfaces, where high scan speed is essential.

NON-CONTACT AFM

Non-contact AFM (NC-AFM) is one of several vibrating cantilever techniques in which an AFM cantilever is vibrated near the surface of a sample. The spacing between the tip and the sample for NC-AFM is on the order of tens to hundreds of angstroms. This spacing is indicated on the van der Waals curve of Figure as the non-contact regime.

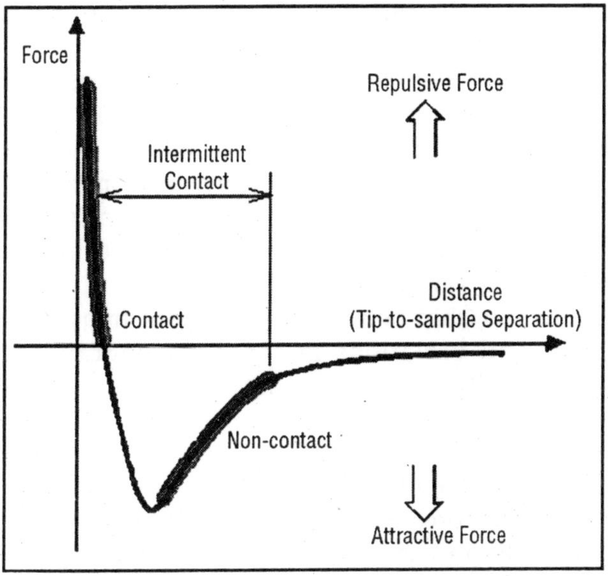

Fig. Interatomic Force vs. Distance Curve

NC-AFM is desirable because it provides a means for measuring sample topography with little or no contact between the tip and the sample. Like contact AFM, non-contact AFM can be used to measure the topography of insulators and semiconductors as well as electrical conductors. The total force between the tip and the sample in the non-contact regime is very low, generally about 10-12N. This low force is advantageous for studying soft or elastic samples. A further advantage is that samples like silicon wafers are not contaminated through contact with the tip.

Because the force between the tip and the sample in the non-contact regime is low, it is more difficult to measure than the force in the contact regime, which can be several orders of magnitude greater. In addition, cantilevers used for NC-AFM must be stiffer than those used for contact AFM because soft cantilevers can be pulled into contact with the sample surface. The small force values in the non-contact regime and the greater stiffness of the cantilevers used for NC-AFM are both factors that make the NC-AFM signal small, and therefore difficult to measure. Thus, a sensitive, AC detection scheme is used for NC-AFM operation.

In non-contact mode, the system vibrates a stiff cantilever near its resonant frequency (typically from 100 to 400 kHz) with an amplitude of a few tens of angstroms. Then it detects changes in the resonant frequency or vibration amplitude as the tip comes near the sample surface. The sensitivity of this detection scheme provides sub-angstrom vertical resolution in the image, as with contact AFM.

The relationship between the resonant frequency of the cantilever and variations in sample topography can be explained as follows. The resonant frequency of a cantilever varies as the square root of its spring constant. In addition, the spring constant of the cantilever varies with the force gradient experienced by the cantilever.

Finally, the force gradient, which is the derivative of the force versus distance curve shown, changes with tip-to-sample separation. Thus, changes in the resonant frequency of a cantilever can be used as a measure of changes in the force gradient, which reflect changes in the tip-to-sample spacing, or sample topography.

In NC-AFM mode, the system monitors the resonant frequency or vibrational amplitude of the cantilever and keeps it constant with the aid of a feedback system that moves the scanner up and down. By keeping the resonant frequency or amplitude constant, the system also keeps the average tip-to-sample distance constant. As with contact AFM (in constant-force mode), the motion of the scanner is used to generate the data set.

NC-AFM does not suffer from the tip or sample degradation effects that are sometimes observed after taking numerous scans with contact AFM. As mentioned above, NC-AFM is also preferable to contact AFM for measuring soft samples. In the case of rigid samples, contact and non-contact images may look the same.

However, if a few monolayers of condensed water are

lying on the surface of a rigid sample, for instance, the images may look quite different. An AFM operating in contact mode will penetrate the liquid layer to image the underlying surface, whereas in non-contact mode an AFM will image the surface of the liquid layer.

For cases where a sample of low moduli may be damaged by the dragging of an AFM tip across its surface, another mode of AFM operation is available: intermittent-contact mode. Intermittent-contact mode is useful for a variety of applications.

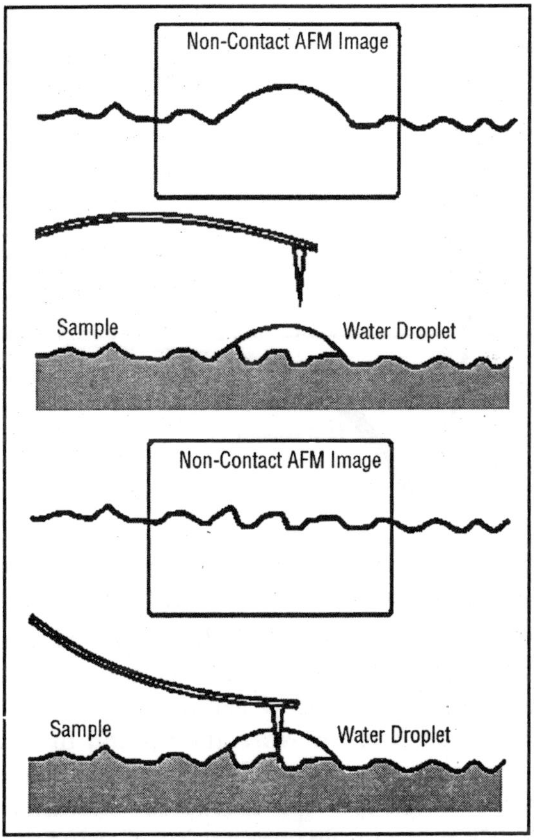

Fig. Contact and Non-Contact AFM images of a Surface with a Droplet of Water

Intermittent-Contact AFM

Intermittent-contact atomic force microscopy (IC-AFM) is similar to NC-AFM, except that for IC-AFM the vibrating cantilever tip is brought closer to the sample so that at the bottom of its travel it just barely hits, or "taps" the sample. The IC-AFM operating region is indicated on the van der Waals curve. As for NC-AFM, for IC-AFM the cantilever's oscillation amplitude changes in response to tip-to-sample spacing. An image representing surface topography is obtained by monitoring these changes.

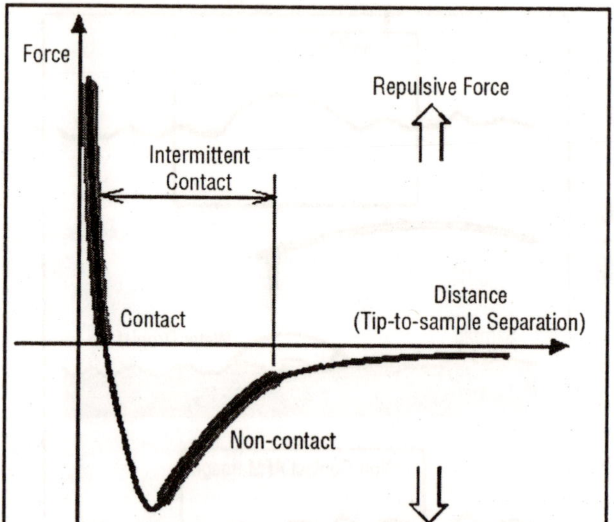

Fig. Interatomic Force vs. Distance Curve

Some samples are best handled using IC-AFM instead of contact or non-contact AFM. IC-AFM is less likely to damage the sample than contact AFM because it eliminates lateral forces (friction or drag) between the tip and the sample. In general, it has been found that IC-AFM is more effective than NC-AFM for imaging larger scan sizes that may include greater variation in sample topography. IC-AFM has become an important AFM technique since it overcomes some of the limitations of both contact and non-contact AFM.

Magnetic Force Microscopy

Magnetic force microscopy (MFM) images the spatial variation of magnetic forces on a sample surface. For MFM, the tip is coated with a ferromagnetic thin film. The system operates in non-contact mode, detecting changes in the resonant frequency of the cantilever induced by the magnetic field's dependence on tip-to-sample separation. MFM can be used to image naturally occurring and deliberately written domain structures in magnetic materials.

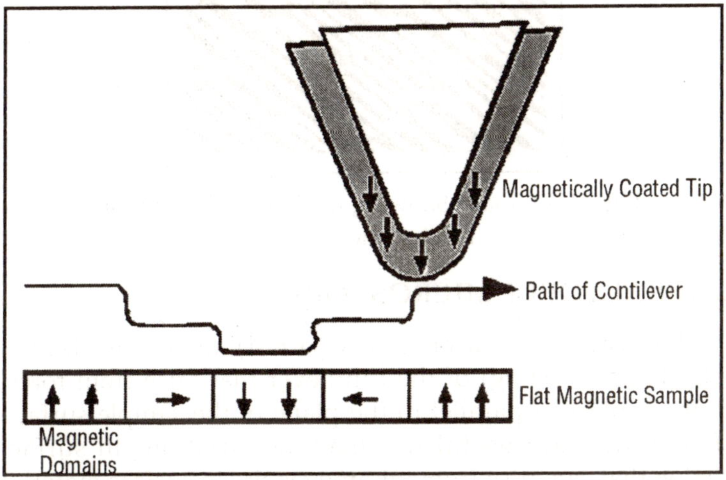

Fig. MFM Maps the Magnetic Domains of the Sample Surface

An image taken with a magnetic tip contains information about both the topography and the magnetic properties of a surface.

Which effect dominates depends upon the distance of the tip from the surface, because the interatomic magnetic force persists for greater tip-to-sample separations than the van der Waals force. If the tip is close to the surface, in the region where standard non-contact AFM is operated, the image will be predominantly topographic.

As we increase the separation between the tip and the sample, magnetic effects become apparent. Collecting a series of images at different tip heights is one way to separate magnetic from topographic effects.

An image of a hard disk acquired in MFM mode is shown in Figure.

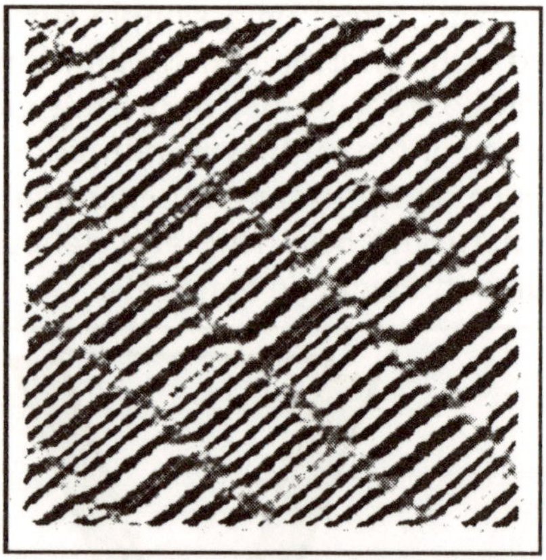

Fig. MFM Image Showing the Bits of a Hard Disk. Field of View 30 μm

LATERAL FORCE MICROSCOPY

Lateral force microscopy (LFM) measures lateral deflections (twisting) of the cantilever that arise from forces on the cantilever parallel to the plane of the sample surface. LFM studies are useful for imaging variations in surface friction that can arise from inhomogeneity in surface material, and also for obtaining edge-enhanced images of any surface.

Lateral deflections of the cantilever usually arise from two sources: changes in surface friction and changes in slope. In the first case, the tip may experience greater friction as it traverses some areas, causing the cantilever to twist more strongly. In the second case, the cantilever may twist when it encounters a steep slope. To separate one effect from the other, LFM and AFM images should be collected simultaneously.

LFM uses a position-sensitive photodetector to detect the deflection of the cantilever, just as for AFM. The difference is that for LFM, the PSPD also senses the cantilever's twist, or lateral deflection. Figure illustrates the difference between an AFM measurement of the *vertical* deflection of the cantilever, and an LFM measurement of *lateral* deflection.

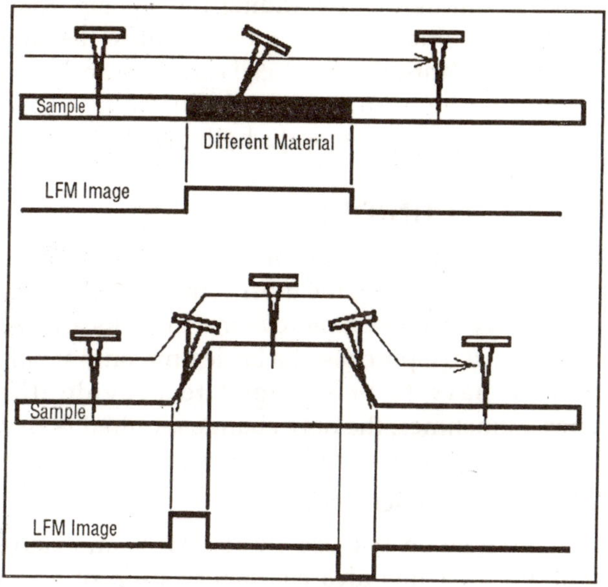

Fig. Lateral Deflection of the Cantilever from Changes in Surface friction (top) and from changes in slope (bottom)

AFM uses a "bi-cell" PSPD, divided into two halves, A and B. LFM requires a "quad-cell" PSPD, divided into four

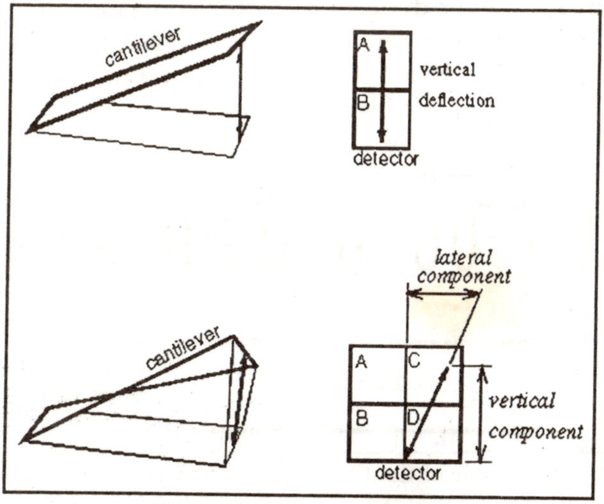

Fig. The PSPD for AFM (top) and LFM (bottom)

quadrants, A through D. By adding the signals from the A and C quadrants, and comparing the result to the sum from the B and D quadrants, the quad-cell can also sense the lateral component of the cantilever's deflection. A properly engineered system can generate both AFM and LFM data simultaneously.

OTHER SPM TECHNIQUES

STM, contact, non-contact, and intermittent-contact AFM, MFM, and LFM are among the most common SPM techniques in practice today. Most commercial instruments are designed so that once we have purchased a basic unit, only small pieces of hardware have to be changed to provide different techniques. Sometimes, the only change required is switching from one mode to another mode using software.

Force Modulation Microscopy

Force modulation microscopy (FMM) is an extension of AFM imaging that includes characterization of a sample's mechanical properties. Like LFM and MFM, FMM allows simultaneous acquisition of both topographic and material-properties data. In FMM mode, the AFM tip is scanned in contact with the sample, and the z feedback loop maintains a constant cantilever deflection (as for constant-force mode AFM). In addition, a periodic signal is applied to either the tip or the sample. The amplitude of cantilever modulation that results from this applied signal varies according to the elastic properties of the sample.

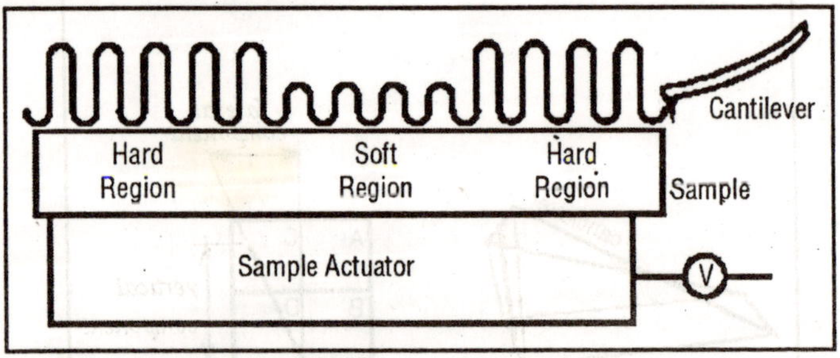

Fig. The Amplitude of Cantilever Oscillation Varies According to the Mechanical Properties of the Sample Surface

The system generates a force modulation image, which is a map of the sample's elastic properties, from the changes in the amplitude of cantilever modulation. The frequency of the applied signal is on the order of hundreds of kilohertz, which is faster than the z feedback loop is set up to track. Thus, topographic information can be separated from local variations in the sample's elastic properties, and the two types of images can be collected simultaneously.

Phase Detection Microscopy

Phase detection microscopy (PDM) - also referred to as phase imaging - is another technique that can be used to map variations in surface properties such as elasticity, adhesion, and friction. Phase detection images can be produced while an instrument is operating in any vibrating cantilever mode, such as non-contact AFM, intermittent-contact AFM (IC-AFM), or MFM mode. Phase detection information can also be collected while a force modulation image is being taken.

Phase detection refers to the monitoring of the phase lag between the signal that drives the cantilever to oscillate and the cantilever oscillation output signal. Changes in the phase lag reflect changes in the mechanical properties of the sample surface.

The system's feedback loop operates in the usual manner, using changes in the cantilever's deflection or vibration amplitude to measure sample topography. The phase lag is monitored while the topographic image is being taken so that images of topography and material properties can be collected simultaneously.

One application of phase detection is to obtain material-properties information for samples whose topography is best measured using IC-AFM rather than contact AFM. For these samples, phase detection is useful as an alternative to force modulation microscopy, which uses contact AFM to measure topography.

Figure shows a topographic non-contact AFM image (left) and a PDM image (right) of an adhesive label. The PDM image provides complementary information to the topography image, revealing the variations in the surface properties of the adhesive label.

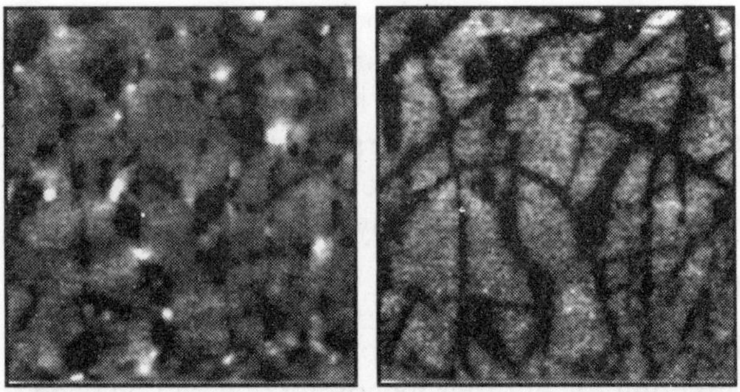

Fig. Non-Contact AFM image (left) and PDM image (right) of an Adhesive Label, collected simultaneously. Field of view 3 μm

Electrostatic Force Microscopy

Electrostatic force microscopy (EFM) applies a voltage between the tip and the sample while the cantilever hovers above the surface, not touching it. The cantilever deflects when it scans over static charges.

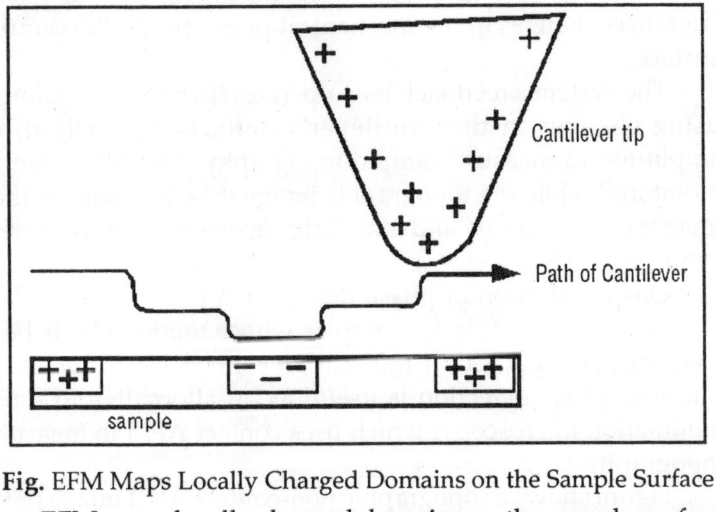

Fig. EFM Maps Locally Charged Domains on the Sample Surface

EFM maps locally charged domains on the sample surface, similar to how MFM plots the magnetic domains of the sample surface. The magnitude of the deflection, proportional to the charge density, can be measured with the standard beam-bounce system. EFM is used to study the spatial variation of surface charge carrier density. For instance, EFM can map the electrostatic fields of a electronic circuit as the device is turned on and off. This technique is known as "voltage probing" and is a valuable tool for testing live microprocessor chips at the sub-micron scale.

Scanning Capacitance Microscopy

Scanning capacitance microscopy (SCM) images spatial variations in capacitance. Like EFM, SCM induces a voltage between the tip and the sample. The cantilever operates in non-contact, constant-height mode.

A special circuit monitors the capacitance between the tip and the sample. Since the capacitance depends on the dielectric constant of the medium between the tip and sample, SCM studies can image variations in the thickness of a dielectric material on a semiconductor substrate. SCM can also be used to visualize sub-surface charge-carrier distributions. For example, to map dopant profiles in ion-implanted semiconductors.

THERMAL SCANNING MICROSCOPY

Thermal scanning microscopy (TSM) measures the thermal conductivity of the sample surface with the tip and the sample not in contact. Like MFM, LFM, and EFM, TSM allows simultaneous acquisition of both topographical and thermal-conductivity data.

For TSM, a cantilever composed of two different metals is used. (Or, a thermal element made up of two metal wires can be used.) The materials of the catilever resopond differently to changes in thermal conductivity, and cause the cantilever to deflect.

The system generates a TSM image, which is a map of the thermal conductivity, from the changes in the deflection of the cantilever. A topographical non-contact image can be generated from changes in the cantilever's amplitude of vibration. Thus, topographic information can be seperated from local variations in the sample's thermal properties, and the two types of images can be collected simultaneously.

NEAR-FIELD SCANNING OPTICAL MICROSCOPY

Near-field scanning optical microscopy (NSOM) uses a special type of scanning probe microscope that operates using visible light. Traditionally, the resolution of optical microscopes has been limited by the wavelength of light to about half a micron. NSOM improves the resolution of an optical microscope by an order of magnitude.

The probe for NSOM is a "light funnel" scanned over the sample. Visible light emanates from the narrow end of the light funnel, a couple of hundred angstroms in diameter, and either reflects off the sample or travels through it into a detector. The intensity of the optical signal collected by the detector at each measurement point constitutes the data set, the NSOM image. NSOM can be used to generate a visible-light image of the surface with a resolution of about 150Å, provided that the distance between the light source and the sample is very short, about 50Å.

NANOLITHOGRAPHY

Fig. Image of a Photoresistive Surface that has Been Modified using Nanolithography. Field of view 40 μm

Figure shows a photoresistive surface that has been modified using this technique. Normally an SPM is used to image a surface without damaging it in any way. However, either an AFM or STM can be used to modify the surface deliberately, by applying either excessive force with an AFM, or high-field pulses with an STM. Not only scientific literature, but also newspapers and magazines have shown examples of surfaces that have been modified atom by atom. This technique is known as nanolithography.

SPMS AS SURFACE ANALYSIS TOOLS

SPMs are most commonly thought of as tools for generating images of a sample's surface. SPMs can also be used, however, to measure material properties at a single x,y point on the sample's surface.

An STM can be used as a spectroscopy tool, probing the electronic properties of a material with atomic resolution. By comparison, the more traditional spectroscopies like XPS (x-ray photoelectron spectroscopy), UPS (ultra-violet photoelectron spectroscopy), or IPES (inverse photo-emission spectroscopy), detect and average the data originating from a relatively large area, a few microns to a few millimeters across. The study of the dependence of an STM signal upon the local electronic structure of the surface is known as scanning tunneling spectroscopy (STS).

The analogy for an AFM is the measurement of force vs. distance curves, which provide information about the local elastic properties of a surface. A force vs. distance curve is a plot of the force on an AFM cantilever tip in the z direction as a function of the z position of the piezoelectric scanner tube.

SCANNING TUNNELING SPECTROSCOPY

Scanning tunneling spectroscopy (STS) studies the local electronic structure of a sample's surface. The electronic structure of an atom depends upon its atomic species (whether it is a gallium atom or an arsenic atom, for instance) and also upon its local chemical environment (how many neighbors it has, what kind of atoms they are, and the symmetry of their distribution). STS encompasses many methods: taking "topographic" (constant-current) images using different bias voltages and comparing them; taking current (constant-height) images at different heights; and ramping the bias voltage with the tip positioned over a feature of interest while recording the tunneling current. The last example results in current vs. voltage (I-V) curves characteristic of theelectronic structure at a specific x,y location on the sample surface. STMs can be set up to collect I-V curves at every point in a data set, providing a three-dimensional map of electronic structure. With a lock-in amplifier, dI/dV (conductivity) or dI/dz (work function) vs. V curves can be collected directly. All of these are ways of probing the local electronic structure of a surface using an STM.

Force vs. Distance Curves

Force vs. distance (F vs. d) curves are used to measure the vertical force that the tip applies to the surface while a contact-AFM image is being taken. This technique can also be used to analyse surface contaminants' viscosity, lubrication thickness, and local variations in the elastic properties of the surface.

Strictly speaking, a force vs. distance curve is a plot of the deflection of the cantilever versus the extension of the piezoelectric scanner, measured using a position-sensitive photodetector. The van der Waals force curve represents just one contribution to the cantilever deflection. Local variations in the form of the F vs. d curve indicate variations in the local elastic properties. Contaminants and lubricants affect the measurement, as does the thin layer of water that is often present when operating an AFM in air.

In the laboratory, force vs. distance curves are quite complex and specific to the given system under study. The discussion here refers to Figure and represents a gross simplification, where shapes, sizes, and distances should not be taken literally.

With this in mind, consider the simplest case of AFM in vacuum. At the left side of the curve, the scanner is fully retracted and the cantilever is undeflected since the tip is not touching the sample. As the scanner extends, the cantilever remains undeflected until it comes close enough to the sample surface for the tip to experience the attractive van der Waals force. The tip snaps into the surface. Equivalently, the cantilever suddenly bends slightly towards the surface.

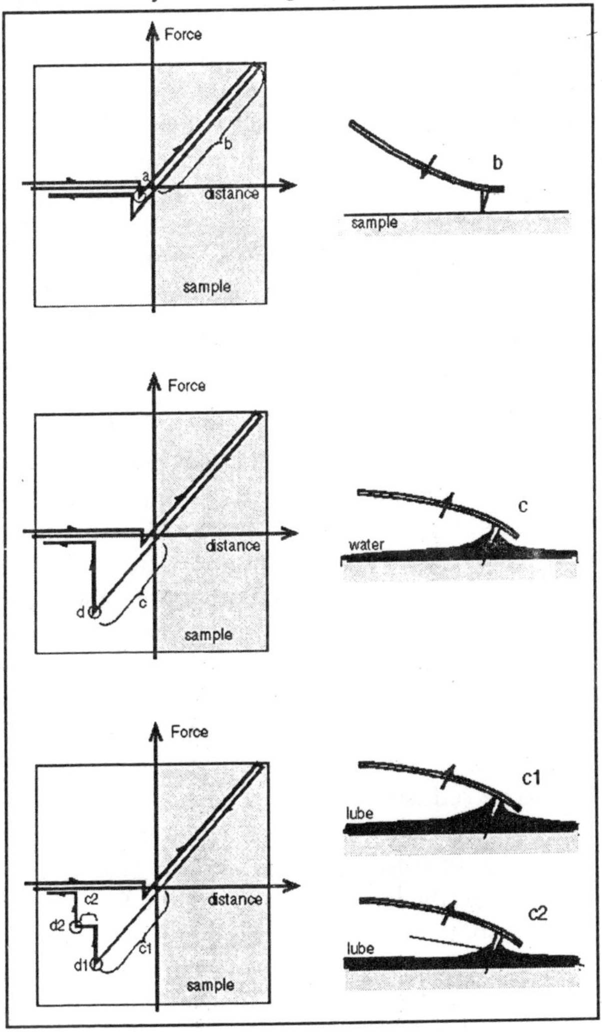

Fig. Forces vs. Distance Curves in Vacuum (top), Air (middle), and Air with a Contamination Layer (bottom)

As the scanner continues to extend, the cantilever deflects

away from the surface, approximately linearly. After full extension, at the extreme right of the plot, the scanner begins to retract. The cantilever deflection retraces the same curve (in the absence of scanner hysteresis) as the scanner pulls the tip away from the surface.

In air, the retracting curve is often different because a monolayer or a few monolayers of water are present on many surfaces. This water layer exerts a capillary force that is very strong and attractive. As the scanner pulls away from the surface, the water holds the tip in contact with the surface, bending the cantilever strongly towards the surface. At some point, depending upon the thickness of the water layer, the scanner retracts enough that the tip springs free. This is known as the snap-back point. As the scanner continues to retract beyond the snap-back point, the cantilever remains undeflected as the scanner moves it away from the surface in free space.

If a lubrication layer is present along with the water layer, multiple snap-back points can occur, bottom. The positions and amplitudes of the snap-back points depend upon the viscosity and thickness of the layers present on the surface.

Contact AFM can be operated anywhere along the linear portion of the force vs. distance curves, in regions (b) or (c). Operation in region (c) might be used for soft samples, to minimize the total force between the tip and the sample. Operating with the cantilever bent towards the surface is inherently a less stable situation, and maximum scan speeds may have to be reduced.

Note that operation in region (c) is still called contact mode, since the tip is touching the sample. Non-contact AFM is operated just to the left of point (a) on the F vs. d curve-just before the tip snaps into the surface.

In the linear region of the F vs. d curves (b), the slope is related to the elastic modulus of the system. When the cantilever is much softer than the sample surface, as is the case for non-destructive imaging, the slope of the curve mostly reflects the spring constant of the cantilever. However, when the cantilever is much stiffer than the sample surface, the slope of the F vs. d curve allows investigation of the elastic properties of the sample.

SPM ENVIRONMENTS

ULTRA-HIGH VACUUM

The first STMs were operated primarily in UHV to study atomically clean surfaces. Silicon has been the most extensively studied material, and obtaining images of the 7 x 7 surface reconstruction of Si (111) is often used as a benchmark for evaluating the performance of UHV STM. A major application of UHV STM is scanning tunneling spectroscopy. STS applied to atomically clean surfaces allows characterization of both topographic and electronic structure, without the added complication of contamination that is always present in air. Another application is the study of materials processes in situ-that is, in the same vacuum environment in which the materials are grown. In this way, processing may proceed without contaminating the sample.

Although the first scanning probe microscope was a UHV STM, the development of UHV AFM came much later. Changing cantilevers and aligning the beam-bounce detection scheme - procedures that are not too difficult with an ambient AFM - were major obstacles to the development of a UHV AFM. However, the operation of both STM and AFM on the same vacuum flange is possible using piezoresistive cantilevers. A combined UHV STM/AFM enables scientists to work with conducting or non-conducting samples.

AMBIENT

The easiest, least expensive, and thus most popular environment for SPMs is ambient, or air. STM in air is difficult, since most surfaces develop a layer of oxides or other contaminants that interfere with the tunneling current. One class of materials for which ambient STM works well is layered compounds.

In graphite, MoS2, Nb3Se, and so forth, a clean, "fresh" surface can be prepared by peeling away older surfaces. AFM is indifferent to sample conductivity, and thus can image any solid surface in air, including the contaminants.

LIQUID

Liquid cells for the SPM allow operation with the tip and the sample fully submerged in liquid, providing the capability for imaging hydrated samples. Operation in liquid can also reduce the total force that the tip exerts on the sample, since the large capillary force is isotropic in liquid.

A liquid environment is useful for a variety of SPM applications, including studies of biology, geologic systems, corrosion, or any surface study where a solid-liquid interface is involved. AFM, LFM, and IC-AFM, FMM, and PDM images can be taken in a liquid environment. A IC-AFM image of rabbit muscle myofibril taken in liquid.

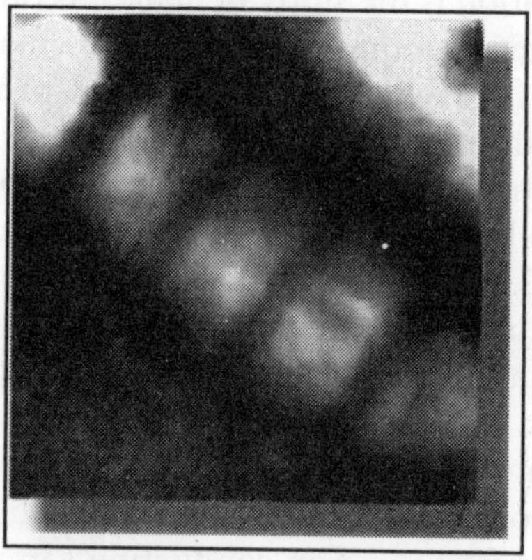

Fig. IC-AFM Image of Rabbit Muscle Myofibril taken in Liquid. Field of View 7 μm

ELECTROCHEMICAL

Like UHV, electrochemical cells provide a controlled environment for SPM operation. Usually provided as an option for ambient SPMs, EC-SPMs consist of a cell, a potentiostat, and software. Applications of EC-SPM include real-space imaging of electronic and structural properties of electrodes, including changes induced by chemical and electrochemical processes, phase formation, adsorption, and corrosion as well as deposition of organic and biological molecules in electrolytic solution.

THE SCANNER

In virtually all scanning probe microscopes, a piezoelectric scanner is used as an extremely fine positioning stage to move the probe over the sample (or the sample under the probe). The SPM electronics drive the scanner in a type of raster pattern.

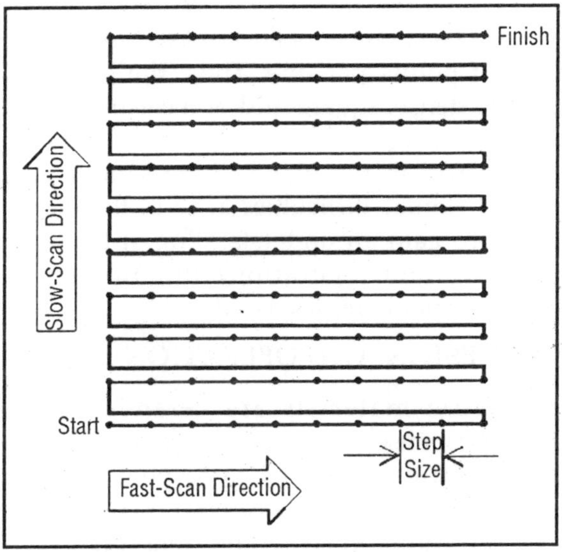

Fig. Scanner motion During data Acquisition

The scanner moves across the first line of the scan, and back. It then steps in the perpendicular direction to the second scan line, moves across it and back, then to the third line, and so forth. The path differs from a traditional raster pattern in that the alternating lines of data are not taken in opposite directions. SPM data are collected in only one direction - commonly called the fast-scan direction - to minimize line-to-line registration errors that result from scanner hysteresis. The perpendicular direction, in which the scanner steps from line to line, is called the slow-scan direction.

While the scanner is moving across a scan line, the image data are sampled digitally at equally spaced intervals. The data are the height of the scanner in z for constant-force mode (AFM) or constant-current mode (STM). For constant-height mode, the data are the cantilever deflection (AFM) or tunneling current (STM).

The spacing between the data points is called the step size. The step size is determined by the full scan size and the number of data points per line. In a typical SPM, scan sizes run from tens of angstroms to over 100 microns, and from 64 to 512 data points per line. (Some systems offer 1024 data points per line.) The number of lines in a data set usually equals the number of points per line. Thus, the ideal data set is comprised of a dense, square grid of measurements.

The difficulties of achieving a perfectly square measurement grid in the plane of the sample surface are explored. And similar difficulties arise in the perpendicular (z) direction. To gain an understanding of the mechanisms underlying scanner operation, the fundamentals of piezoelectric scanners are discussed.

SCANNER DESIGN AND OPERATION

Piezoelectric materials are ceramics that change dimensions in response to an applied voltage. Conversely, they develop an electrical potential in response to mechanical pressure. Piezoelectric scanners can be designed to move in x, y, and z by expanding in some directions and contracting in others. Piezoelectric scanners for SPMs are usually fabricated from lead zirconium titanate, or PZT, with various dopants added to create specific materials properties.

Scanners are made by pressing together a powder, then sintering the material. The result is a polycrystalline solid. Each of the crystals in a piezoelectric material has its own electric dipole moment. These dipole moments are the basis of the scanner's ability to move in response to an applied voltage.

After sintering, the dipole moments within the scanner are randomly aligned. If the dipole moments are not aligned, the scanner has almost no ability to move. A process called poling is used to align the dipole moments. During poling the scanners are heated to about 200°C to free the dipoles, and a DC voltage is applied to the scanner. Within a matter of hours most of the dipoles become aligned. At that point, the scanner is cooled to freeze the dipoles into their aligned state. The newly poled scanner can then respond to voltages by extending and contracting.

Occasional use of the scanner will help maintain the scanner's polarization. The voltage applied to enact the scanning motion realigns the stray dipoles that relax into random orientation. If the scanner is not repoled by regular use, a significant fraction of the dipoles will begin to randomize (depolarize or depole) again over a period of weeks. Depoling is accelerated markedly if the scanner is subjected to temperatures above 150°C.

This means that if we want to add a heated stage to our SPM, we must isolate it thermally from the scanner. (The Curie temperature for PZT materials is about 150°C.) Many SPMs use variations of the simple tube design depicted in Figure. In this design, the scanner is a hollow tube.

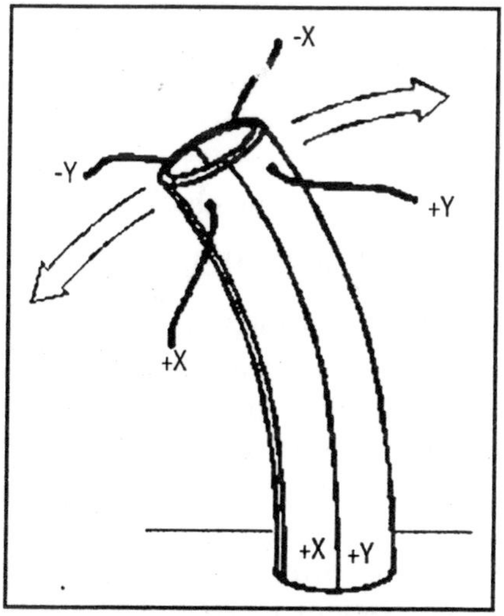

Fig. The Scanner Tube

Electrodes are attached to the outside of the tube,

segmenting it electrically into vertical quarters, for +x, +y, -x, and -y travel. An electrode is also attached to the centre of the tube to provide motion in the z direction. When alternating voltages are applied to the +x and -x electrodes, for example, the induced strain of the tube causes it to bend back and forth in the x direction. Voltages applied to the z electrode cause the scanner to extend or contract vertically.

In most cases, the voltage applied to the z electrode of the scanner at each measurement point constitutes the AFM (constant-force) or the STM (constant-current) data set. In some cases, an external sensor is used to measure the height of the scanner directly). It is assumed that the data set consists of voltages applied to the z electrode.

The maximum scan size that can be achieved with a particular piezoelectric scanner depends upon the length of the scanner tube, the diameter of the tube, its wall thickness, and the strain coefficients of the particular piezoelectric ceramic from which it is fabricated. Typically, SPMs use scanners that can scan laterally from tens of angstroms to over 100 microns. In the vertical direction, SPM scanners can distinguish height variations from the sub-angstrom range to about 10 microns.

Piezoelectric scanners are critical elements in SPMs, valued for their sub-angstrom resolution, their compactness, and their high-speed response. However, along with these essential properties come some challenges.

SCANNER NONLINEARITIES

As a first approximation, the strain in a piezoelectric scanner varies linearly with applied voltage. (Strain is the change in length divided by the original length) The following equation describes the ideal relationship between the strain and an applied electric field:

$$S = d\,E$$

where s is the strain in Å/m, E is the electric field in V/m, and d is the strain coefficient in Å/V. The strain coefficient is characteristic of a given piezoelectric material.

In practice, the behaviour of piezoelectric scanners is not

so simple-the relationship between strain and electric field diverges from ideal linear behaviour.

INTRINSIC NONLINEARITY

Suppose the applied voltages start from zero and gradually increase to some finite value. If the extension of the piezoelectric material is plotted as a function of the applied voltage, the plot is not a straight line, but an s-shaped curve, as seen in Figure. The straight line in the graph is a linear fit to the data.

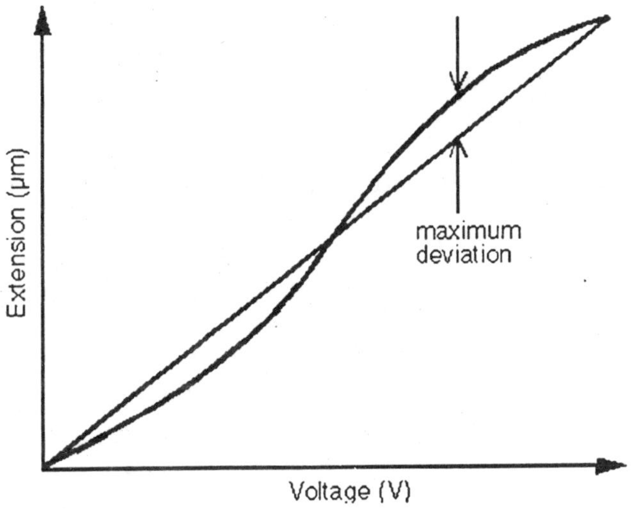

Fig. Intrinsic Nonlinearity of a Scanner

The intrinsic nonlinearity of a piezoelectric material is the ratio of the maximum deviation y from the linear behaviour to the ideal linear extension y at that voltage. In other words, the nonlinearity is the change in y divided by y, expressed as a percentage. Intrinsic nonlinearity typically ranges from 2 per cent to 25 per cent in the piezoelectric materials used in SPM systems.

In the plane of the sample surface, the effect of intrinsic nonlinearity is distortion of the measurement grid. Because the scanner does not move linearly with applied voltage, the measurement points are not equally spaced. As a result, an

SPM image of a surface with periodic structures will show non-uniform spacings and curvature of linear structures. Less regular surfaces may not show recognizable distortion, even though the distortion is present.

Perpendicular to the plane of the sample surface (in the z direction), intrinsic nonlinearity causes errors in height measurements. Height is usually calibrated for SPMs by scanning a sample with a known step height. By reading the voltage applied to the z electrode as it traverses the step, the z component of the strain coefficient can be calculated. However, if that strain coefficient is directly applied to measure a feature different in height from the calibrated height, the intrinsic nonlinearity of the scanner will generate errors.

HYSTERESIS

To complicate matters, piezoelectric ceramics display hysteretic behaviour. Suppose we start at zero applied voltage, gradually increase the voltage to some finite value, and then decrease the voltage back to zero. If we plot the extension of the ceramic as a function of the applied voltage, the descending curve does not retrace the ascending curve - it follows a different path. It is assumed that the voltages change very slowly.

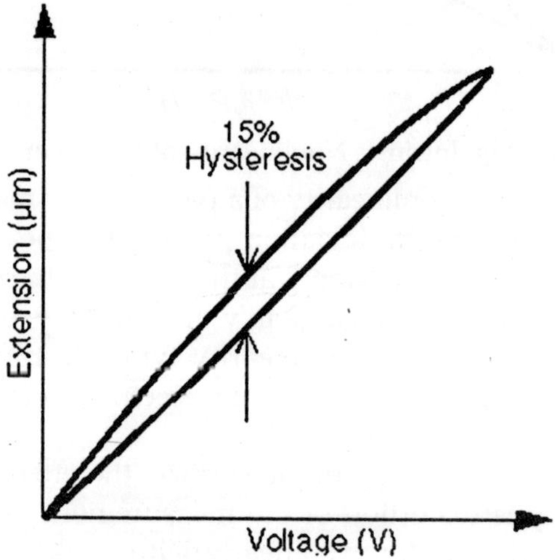

Fig. Hysteresis of a Scanner

The hysteresis of a piezoelectric scanner is the ratio of the maximum divergence between the two curves to the maximum extension that a voltage can create in the scanner (the change in Y/Ymax). Hysteresis can be as high as 20 per cent in piezoelectric materials.

As mentioned SPM data are usually collected in one direction to minimize registration errors caused by scanner hysteresis. Data collected in the return direction would be shifted slightly, distorting the measurement grid. Since most SPMs today have provisions for taking scans in any fast-scan direction, we can observe hysteresis in the plane of the sample by comparing data sets taken in opposite fast-scan directions. For example, we could compare a data set taken left to right with a data set taken right to left. Scanner hysteresis would cause a shift between the data sets.

Hysteresis in the direction perpendicular to the plane of the sample causes erroneous step-height profiles. Referring to the curve of Figure, we see that if the scanner is going up a step in the z direction, a certain voltage is required to allow the scanner to contract. But going down the same step the scanner extends, and extension takes more voltage than contraction for the same displacement. When the SPM image is represented by the voltage applied to the scanner. It is arbitrarily assumed that the strain coefficient in z was calibrated for a contracting scanner.

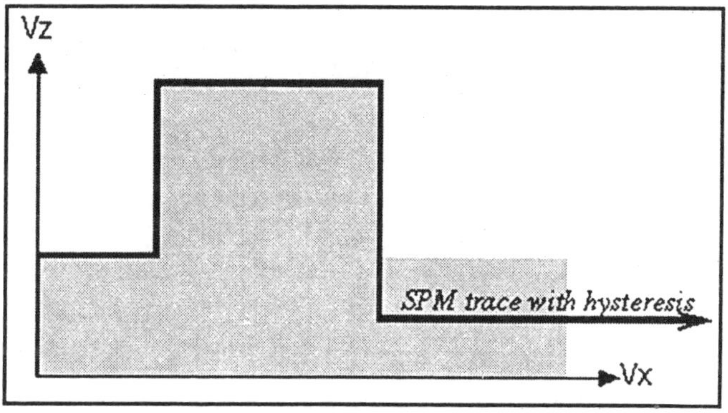

Fig. Effects of Hysteresis on a Step.

CREEP

When an abrupt change in voltage is applied, the piezoelectric material does not change dimensions all at once. Instead, the dimensional change occurs in two steps: the first step takes place in less than a millisecond, the second on a much longer time scale. The second step, delta xc, is known as creep.

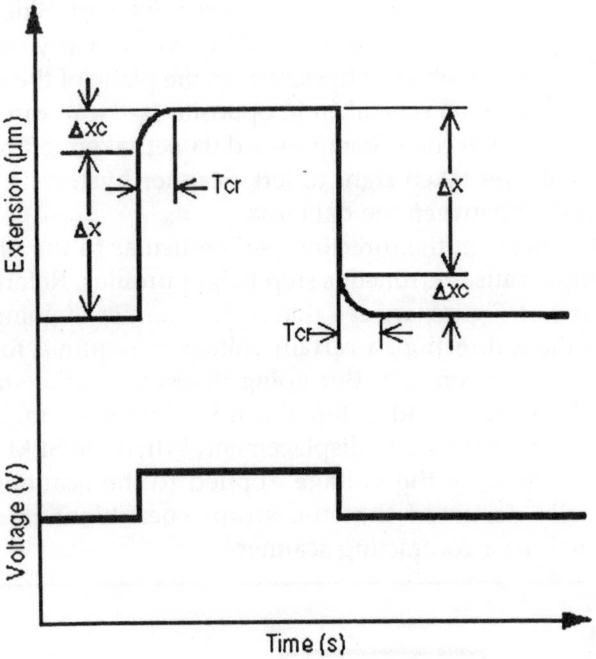

Fig. Plots Showing Creep of a Scanner.

Quantitatively, creep is the ratio of the second dimensional change to the first: the change in xc divided by the change in x. Creep is expressed as a percentage and is usually quoted with the characteristic time interval Tcr over which the creep occurs. Typical values of creep range from 1 per cent to 20 per cent, over times ranging from 10 to 100 seconds.

The times involved during typical scans place the lateral motion of the scanner in the curved part of the response curve.

As a result, two scans taken at different scan speeds show slightly different length scales (magnifications) when creep is present. We can trust only the measurement made at the scan speed that we used to calibrate our SPM when creep is present.

Another effect of creep in the plane of the sample is that it can slow down our work when we try to zoom in on a feature of interest. Suppose we want to characterize a defect on the surface of our sample.

Most likely, we 'll first take a large scan to find the defect. Suppose we see the defect, very tiny, in a corner of the large scan. What we would prefer to do is take a much smaller scan - a high-resolution scan-centered about the defect. An SPM applies an offset voltage to move the scanner to its new centre position to take the second scan.

Because we have applied a sudden voltage step, in this case an offset voltage in the plane of the sample, creep will cause we to miss our target. If we continue to collect scans without changing the offset voltage, the scanner will continue to relax until the defect eventually drifts (creeps) into the centre of our scan. The process will likely take a matter of minutes.

To examine the effects of creep in the z direction. As the tip traverses the step from bottom to top, the scanner contracts immediately with a voltage corresponding to the full step height. However, over the next few seconds the scanner continues to contract slowly as creep occurs.

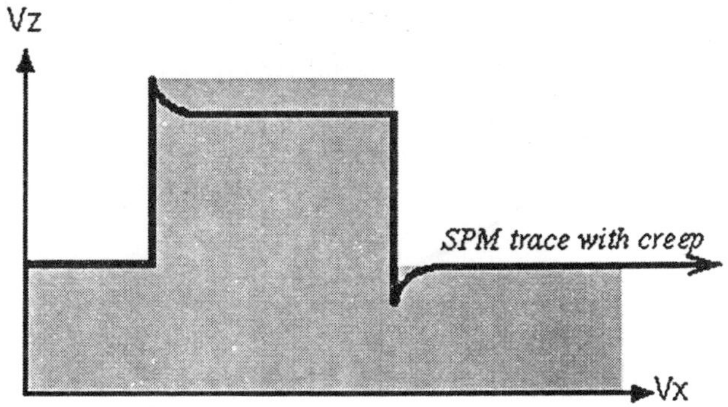

Fig. Effects of Creep on a Step.

To keep the tip in contact with the sample, an SPM has to apply a voltage in the other direction, counteracting the creep. When the tip traverses the step from top to bottom, the same process occurs. The scanner extends to accommodate the step, then continues to creep. An SPM must apply a voltage in the opposite direction to maintain contact between the tip and the sample.

The effect of creep on the profile of a step is a simple example. Consider a more typical sample with steep slopes that represent defects or grains or other surface features. Creep may cause an SPM image to look like it has ridges on one side of a feature and shadows on the other side of a feature. If there are shadows where the tip has just traveled down a steep slope, or ridges where the tip has just traveled up a steep slope, creep may be responsible. Reversing the fast-scan direction and taking the same image helps separate creep artifacts from true ridges and trenches.

AGING

The strain coefficient, d, of piezoelectric materials changes exponentially with time and use. The aging of a piezoelectric scanner, for cases of both high and low voltage usage.

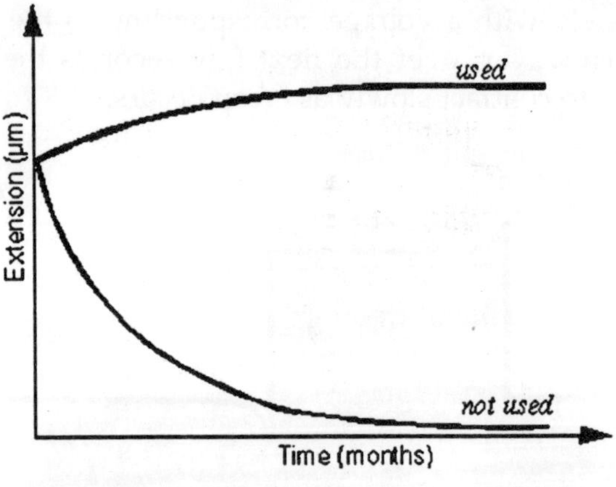

Fig. Aging of a scanner, with use or left idle.

The aging rate is the change in strain coefficient per decade of time. When a scanner is not used, the deflection achieved for a given voltage gradually decreases. The aging rate of scanners for an SPM can produce a decrease in lateral strain coefficients (and therefore, an error in length measurements taken from SPM images) over time. When a scanner is used regularly, the deflection achieved for a given voltage increases slowly with use and time. These two phenomena are part of the same process. Each of the tiny crystals that compose the scanner has its own dipole moment. Repeated application of voltage in the same direction-such as the voltage applied during scanning-causes more and more of the dipoles to align themselves along the axis of the scanner. The amount of deflection achieved for a given voltage depends upon how many dipoles are aligned. Thus, the more the scanner is used, the farther the scanner will travel.

On the other hand, if the scanner is not used, the dipole moments of the crystals will gradually become randomly oriented again. As a result, fewer dipoles contribute to the deflection of the scanner. When we buy an SPM, the scanners have already been "poled," which means that they have already been exercised to the point where the deflection of the scanner is close to its maximum.

The dependence of the scanner deflection on time and usage means that the scanner may not be extending the same distance for a given applied voltage as it did when it was first calibrated. As a result, when we measure a feature on an SPM image, the values of lateral and vertical dimensions may be in error.

CROSS COUPLING

The term cross coupling refers to the tendency of x-axis or y-axis scanner movement to have a spurious z-axis component. It arises from several sources and is fairly complex. For example, the electric field is not uniform across the scanner. The strain fields are not simple constants, but actually complex tensors. Some "cross talk" occurs between x, y, and z electrodes. But the largest effect is geometric. Geometric cross coupling has its basis in the way piezoelectric scanners are constructed, usually as segmented tubes or as tripods.

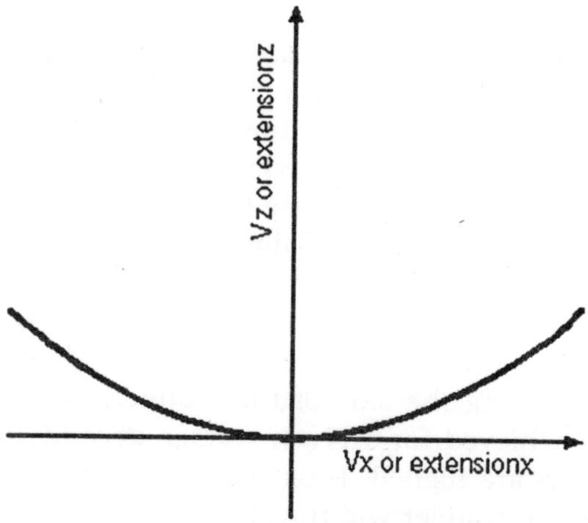

Fig. Cross Coupling Motion of the Scanner

The *x-y* motion of the scanner tube is produced when one side of the tube shrinks and the other side expands. As a result, a piezoelectric tube scans in an arc, not in a plane. A voltage applied to move the piezoelectric tube along the x or y axis (parallel to the surface of the sample) necessitates that the scanner extend and contract along the z axis (perpendicular to the surface of the sample) to keep the tip in contact with the sample.

A tripod scanner is designed with three mutually perpendicular bars or tubes glued together at one end. This design is also susceptible to cross coupling because the three bars of piezoelectric material are attached to one another. When the x bar extends or contracts, it causes rotation of the y and z bars. Cross coupling can cause an SPM to generate a bowl-shaped image of a flat sample. A profile of such an image is shown in Figure with an example of a step.

In interpreting, remember that the SPM image is based on the voltage required to compensate for the curvature generated by the arc of the scanner. The bowl shape may not always be evident in the final image because the curved background can be subtracted out using image-processing software. The best way to determine if the scanner is subject to cross coupling is to image a sample with a known radius of curvature, such as a lens. Software corrections can only flatten it or leave it alone. In the first case, we end up with a spuriously flat image, and in the second case the curvature of the scanner will be added to the curvature of the lens. The true curvature of a lens can be measured only when cross coupling is eliminated.

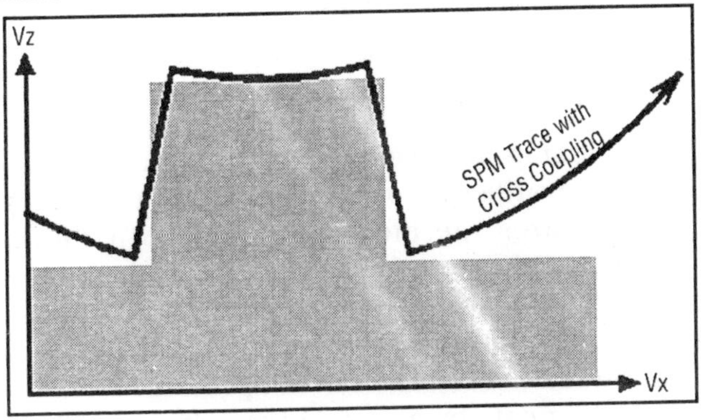

Fig. Effects of Cross Coupling on a Step

The example of a single step to demonstrate hysteresis, creep, and cross coupling in the vertical direction is used. To show a single image in the laboratory that illustrates each of these effects in isolation is virtually impossible. The sum of the effects of hysteresis, creep, and cross coupling in the image of a single step. (The aspect ratio of the tip may also contribute to the shape of the sidewalls)

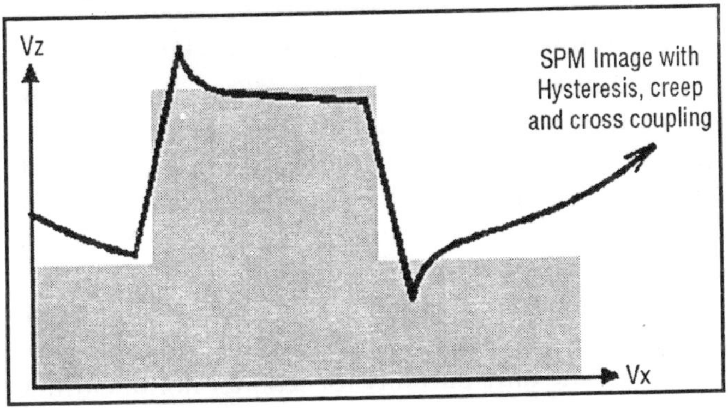

Fig. Effects of Hysteresis, Creep, and Cross Coupling on a Step

Traditionally, the nonlinear behaviour of piezoelectric scanners described above has been addressed imperfectly using software corrections. Some systems on the market use hardware solutions that eliminate most of the nonlinearities instead of correcting them. Hardware solutions are divided into optical, capacitive, and strain-gauge techniques. The best systems combine hardware and software corrections, using the strengths of each.

SOFTWARE CORRECTION

The first step in developing a software correction for scanner nonlinearities is to image a calibration grating. The system compares the measured data with the known characteristics of the grating. Then the system determines how to adjust the measured data to conform to the known characteristics. It stores that information in a file, or lookup table. Afterwards, the system can compensate for nonlinearity while collecting data by adjusting the voltage applied to the scanner in accordance with the file or the look-up table.

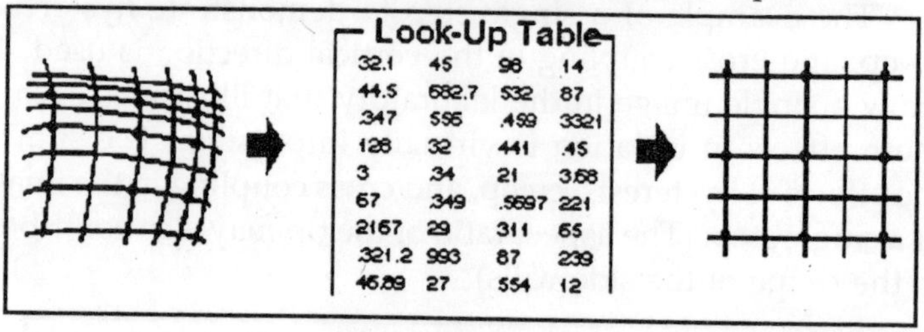

Fig. Software Correction, Shown Using a Look-up table

More sophisticated software correction algorithms include equations that model the scanner tube's non-linear response to applied voltages. The equations are used to calculate the voltages that are applied to the scanner to position it during a scan. Again, scanner calibration procedures are necessary to determine parameters for the algorithm.

Software solutions in general are relatively simple and inexpensive to implement. Their main disadvantage is that they compensate only partially for scanner nonlinearities. The corrections are strongly dependent upon the scan speed, scan direction, and whether the scanner was centered within its scan range during calibration. As a result, software corrections are most accurate only for scans that reproduce the conditions under which the calibration was done.

A scanner should be re-calibrated when scan conditions change. In practice, SPM systems that use only software corrections may vary from true linear behaviour by as much as 10 per cent. Often this deviation is unnoticed by the SPM user who calibrates and checks the calibration of the SPM with the same procedure every time.

HARDWARE CORRECTION

Systems with hardware solutions sense the scanner's actual position with external sensors. The signal read from the sensor of each axis is compared to a signal that represents the intended scanner position along that axis. A feedback system applies voltage to the scanner to drive it to the desired position. In this way, the scanner can be driven in a linear fashion. The external sensors themselves must be stable and immune to all sources of nonlinearity, since their purpose is to improve the linearity of the SPM.

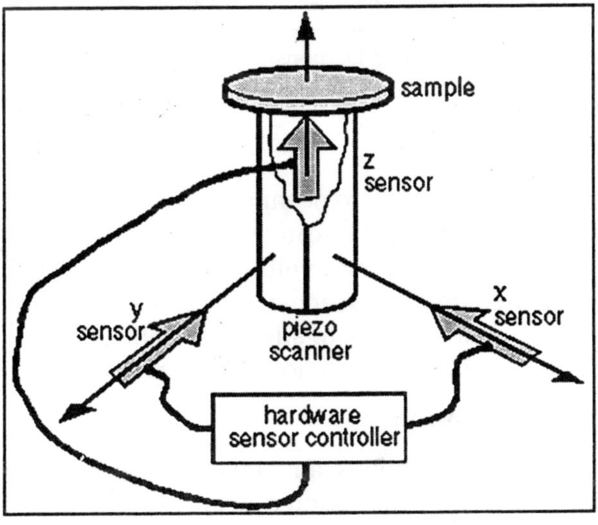

Fig. Schematic of a scanner, showing external detectors

Because the scanner's actual position is measured, systems using hardware sensors compensate for intrinsic nonlinearity, hysteresis, creep, aging, and cross coupling. They can reduce the total nonlinearity of the system to less than 1 per cent.

Figure illustrates how SPMs equipped with hardware correction enable we to zoom in on a feature of interest in an image without the effects of creep. The left image is a 40μm scan of a grating. The box in the image shows the intended location of the zoomed scan. The right image is the zoomed 2μm scan of that location. An image of the same area acquired at a later time would look identical to the right image, indicating that creep is not present in SPMs with hardware correction schemes.

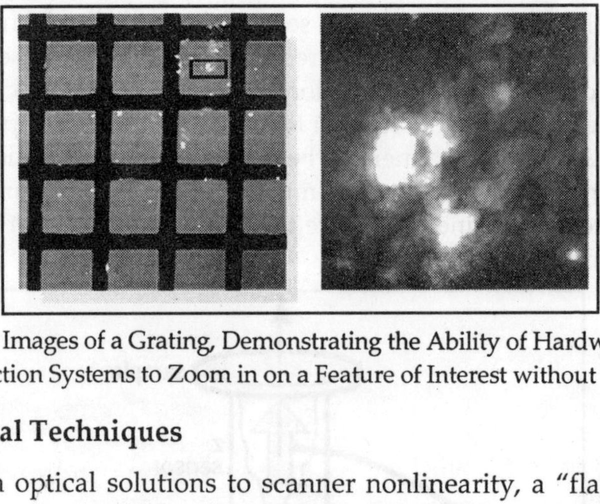

Fig. Images of a Grating, Demonstrating the Ability of Hardware Correction Systems to Zoom in on a Feature of Interest without Creep

Optical Techniques

In optical solutions to scanner nonlinearity, a "flag", or reflector, is attached to the scanner. A light shines on the flag, and a detector responding to either the phase or the position of the light determines the actual position of the scanner. The system then uses feedback to adjust the voltage to the scanner to place it in the proper position to generate a linear scan.

Capacitive Techniques

In capacitive solutions, a metal electrode is placed on the scanner and another is mounted nearby. An electrical circuit then measures the capacitance between the electrodes, which varies with their separation. Having located the scanner in this way, the system uses feedback to adjust the scanner's position to compensate for nonlinearity.

Strain-Gauge Techniques

In strain-gauge solutions to scanner nonlinearities, a strain gauge is mounted on the scanner. The active element in a strain gauge is a piece of piezoresistive material-material whose resistance varies when it is under tensile or compressive stress. When the scanner moves, the resistance in the strain gauge changes. The magnitude of the resistance of the strain gauge reflects the amount of bending in the scanner, thereby measuring its position. An SPM uses feedback to adjust the scanner's position and to generate a linear scan.

Tests for Scanner Linearity

If we suspect that nonlinear behaviour of the piezoelectric scanner has created artifacts in an SPM image, we can perform various tests. The symptoms and gives a test procedure to identify each source of nonlinearity. When we are evaluating an SPM, make sure that it has adequate compensation for intrinsic nonlinearity, hysteresis, creep, aging, and cross coupling. Otherwise, our SPM images may be distorted, the measurements we make from them may be erroneous, and the conclusions we draw from our measurements may lead we astray.

Intrinsic Nonlinearity

X-Y Plane

Symptom: Non-uniform spacings and curvature appear in structures that we know or believe to be linear.

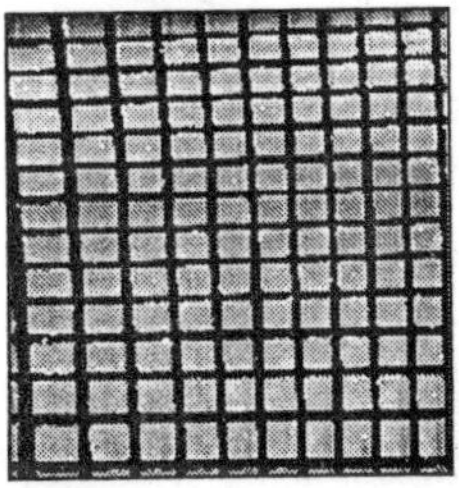

Fig. Image of a Grating, Showing non-uniform Spacing and Curvature

Test: Image a calibration grating. The image should show consistent spacing and straight lines.

Cantilevers

Cantilevers and their tips are critical components of an atomic force microscope system because they determine the force applied to the sample and the ultimate lateral resolution of the system. Figure depicts a scanning electron microscope (SEM) image of a cantilever.

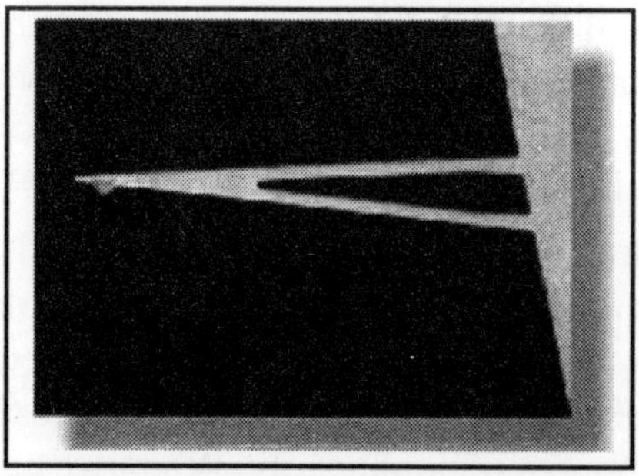

Fig. SEM image of a Cantilever

Integrated tip and cantilever assemblies can be fabricated from silicon or silicon nitride using photolithographic techniques.

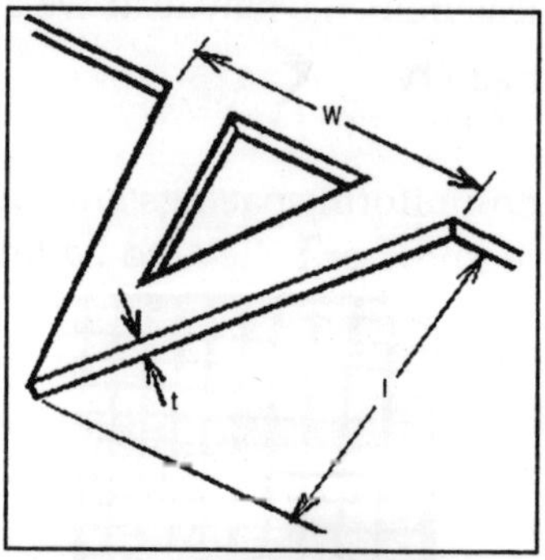

Fig. V-shaped cantilever, showing length (l), width (w), and thickness (t)

More than 1,000 tip and cantilever assemblies can be produced on a single silicon wafer. V-shaped cantilevers are the most popular, providing low mechanical resistance to vertical deflection, and high resistance to lateral torsion. Cantilevers typically range from 100 to 200μm in length, 10 to 40 μm in width, and 0.3 to 2 μm in thickness.

PROPERTIES OF CANTILEVERS

Atomic force microscopes require not only sharp tips, but also cantilevers with optimized spring constants - lower than the spring constants between atoms in a solid, which are on the order of 10 N/m. The spring constant of the cantilever depends on its shape, its dimensions, and the material from which it is fabricated. Thicker and shorter cantilevers tend to be stiffer and have higher resonant frequencies. The spring constants of commercially available cantilevers range over four orders of magnitude, from thousandths of a newton per meter to tens of newtons per meter. Resonant frequencies range from a few kilohertz to hundreds of kilohertz providing high-speed response and allowing for non-contact AFM operation.

HOW TO SELECT A CANTILEVER

The desirable properties for a cantilever depend on the imaging mode and the application. In contact mode, soft cantilevers are preferable because they deflect without deforming the surface of the sample. In non-contact mode, stiff cantilevers with high resonant frequencies give optimal results.

TIP SHAPE AND RESOLUTION

The lateral resolution of an AFM image is determined by two factors: the step size of the image, and the minimum radius of the tip. Consider an image taken with 512 by 512 data points. Such a scan 1 μm by 1μm would have a step size-and lateral resolution-of about 20Å (1 μm ÷ 512).

The sharpest tips available commercially can have a radius as small as 50Å. Because the interaction area between the tip and the sample is a fraction of the tip radius, these tips typically provide a lateral resolution of 10 to 20Å. Thus, the resolution of AFM images larger than 1 μm by 1μm is usually determined not by the tip but by the image's step size. The quoted *best resolution* of an AFM, however, depends upon how resolution is defined.

In the microscopy community, two asperities (peaks) are considered resolved if the image satisfies Rayleigh's criterion. In this application, Rayleigh's criterion requires that the height of the image dip at least 19 per cent between the asperities.

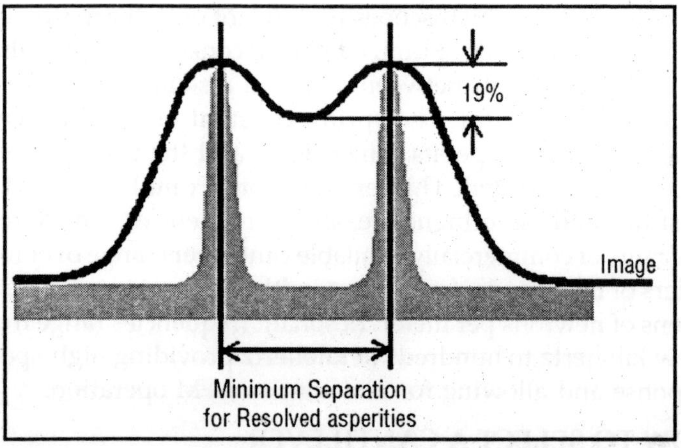

Fig. Definition of Lateral Resolution using Rayleigh's Criterion

To determine the lateral resolution of an SPM experimentally, the asperities are brought closer and closer together until the image no longer dips by 19 per cent between peaks. The minimum separation between resolved asperities determines the best lateral resolution of the system. Using this definition, the lateral resolution of an AFM with the sharpest tips commercially available is 10 to 20Å.

At first glance the figure of 10 to 20Å resolution seems to conflict with the ubiquitous images of atomic lattices in AFM brochures. The distinction between *imaging atomic-scale features* with accurate lattice spacing and symmetry, and *true atomic resolution* bears some comment.

An STM gives true atomic resolution. Because the dependence of the tunneling current on the tip-to-sample separation is exponential, only the closest atom on a good STM tip interacts with the closest atom on the sample, as shown in Figure, top. For AFMs, the dependence of cantilever deflection on tip-to-sample separation is weaker. The result is that several atoms on the tip interact simultaneously with several atoms on the sample.

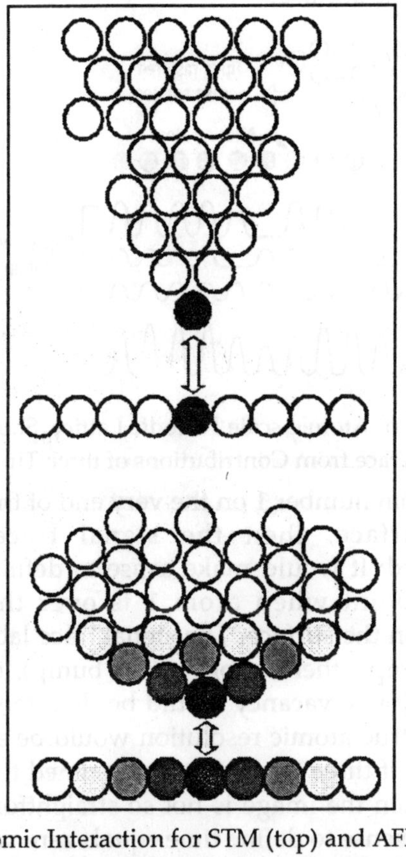

Fig. Interatomic Interaction for STM (top) and AFM (bottom). Shading shows Interaction Strength

Figure illustrates how this multiple interaction affects AFM images: AFMs cannot achieve the true atomic resolution needed to detect, for example, the atomic vacancy depicted in the figure.

The atomic vacancy is shifted in position and not well resolved in the sum signal trace. In the figure, three tip atoms, numbered 1 through 3, are assumed to interact with the atoms of the surface. Three signal traces are also shown, representing the contribution from each atom. The traces show signals varying as a function of time, where the signal reflects the interatomic force between the given tip atom and the atom beneath it on the surface.

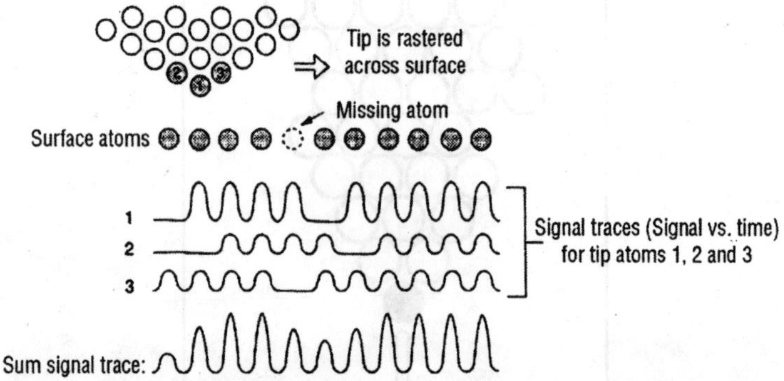

Fig. AFM Scan of Atomic-scale Periodic Lattice, Showing the Sum Signal Trace from Contributions of three Tip Atoms

If only atom number 1 on the very end of the tip interacts with the surface, then the signal trace would be straightforward: It would make sense to define time equals zero at the point when atom 1 is over the first atom encountered on the surface. The signal trace labeled 1 would show four bumps, then a hole (or no bump), then six more bumps. The atomic vacancy would be detected in its correct position, and true atomic resolution would be achieved.

However, if three tip atoms are assumed to interact with the surface, then the image is not so straightforward. In this case, it makes sense to define time equals zero when atom 3 is over the first atom encountered on the surface. When atom 3 encounters the first surface atom, some signal is registered, apparent from the signal trace of atom 3.

At the next instant in time (assuming a step size on the order of one atomic spacing), atom 1 detects the first surface atom, atom 3 detects the second surface atom, and atom 2 still registers no signal. At the third instant in time, all three tip atoms register signals which vary in intensity according to their distance from the surface.

This example shows that, since the tip atoms are displaced laterally, they produce signal traces that are offset from one and other. The vacancy appears in a different location for each signal trace. An image is the result of the sum of the contributions from all interacting tip atoms. Because the three interacting tip atoms are displaced from one and other laterally, there is no clear vacancy apparent in the sum signal. A minimum is apparent, but its location is shifted.

In this simple case, only three tip atoms were assumed to interact with the surface. In reality, the number of interacting tip atoms for contact AFM is large, and at each instant in time their interaction is with several surface atoms. For larger numbers of interacting atoms, features such as vacancies are even less well-defined and more shifted. Thus, while the periodicity of the lattice is reproduced, true atomic resolution is not achieved.

For both contact and non-contact imaging, we should choose a tip that is sharper than the smallest features on our sample. If the tip is larger than the surface features, an artifact know as *tip imaging* occurs. We may not always want to choose the sharpest tip available, since the sharpest tips are more expensive and can be less durable. We need the sharpest tips only when we require the best resolution.

For LFM, blunter tips can be preferable precisely because they present a larger interaction area between tip and sample. A larger area of interaction can produce a stronger lateral deflection of the cantilever. The larger cantilever deflection must be balanced against the lower lateral resolution delivered by the blunter tip.

AFM manufacturers offer microfabricated tips in three geometries: pyramidal, tetrahedral, and conical. Conical tips can be made sharp, with high aspect ratios (the ratio of tip length to tip width). Tip radii as small as 50Å have been observed. Pyramidal tips have lower aspect ratios and nominal tip radii of a few hundred angstroms, but they are more durable.

AFM tips are fabricated from silicon or silicon nitride. The fabrication process is different for the two materials. The features of tips made from each material are governed by the fabrication process as well as the material properties.

Silicon conical tips are made by etching into the silicon around a silicon dioxide cap. The high aspect ratio of conical tips makes them suitable for imaging deep, narrow features such as trenches, but they may break more easily than the pyramidal or tetrahedral geometries, as described above. Silicon also has the advantage that it can be doped, so that tips can be made electrically conducting.

Conducting tips are useful for controlling the bias between the tip and the sample, or for preventing unwanted charge buildup on the tip.

Silicon nitride tips are fabricated by depositing a layer of silicon nitride over an etched pit in a crystalline silicon surface. This method produces the pyramidal or tetrahedral tip geometry. The aspect ratio of a silicon nitride tip is thus limited by the crystallographic structure of the etch pit material, silicon. The tips are broader than conical silicon tips, making them sturdier but less suitable for imaging deep, narrow features. Silicon nitride is a harder material than silicon, which also makes silicon nitride tips more durable than silicon tips.

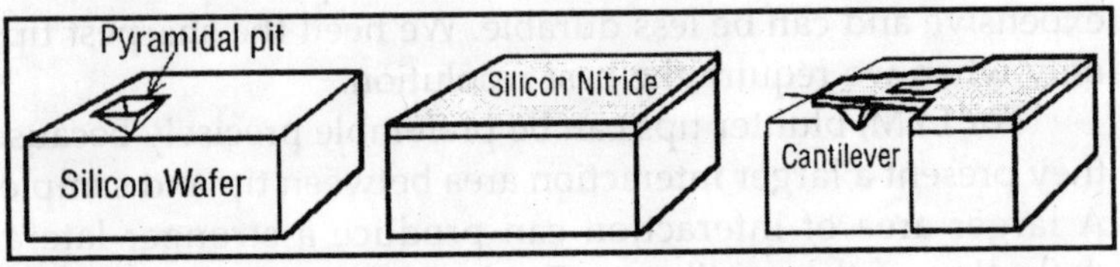

Fig. Fabrication of Silicon Nitride Tip

Silicon nitride films, however, contain residual stresses which make them deform as the film thickness increases. For this reason, applications that call for thick cantilevers - or cantilevers with high resonant frequencies - use silicon cantilevers. The thickness of silicon nitride cantilevers is usually less than one micron, whereas silicon cantilevers can be several microns thick.

A more specialized type of cantilever that has a tip grown in a scanning electron microscope is also available commercially. (A specialized tip can also be produced by "milling" an existing tip using a focused ion beam.) The SEM is used in an unusual way: to deposit contamination in a column. The tips are not microfabricated in bulk; they are grown one at a time on top of a bare cantilever or square-pyramidal tip. These tips have the advantage of almost unlimited aspect ratio. However, they are not normally sharp, and can bend or break easily.

IMAGE ARTIFACTS

SPM images are among the easiest to interpret of images generated by any microscopy technique. With an electron or optical microscope, contrast is based on complex electromagnetic diffraction effects. Determining whether a feature is protruding from the surface or recessed into it can be difficult with an image from an optical or electron microscope. SPMs, however, collect three-dimensional data. In an SPM image, a peak is unambiguously a peak, and a valley is clearly a valley. With an optical or electron microscope, artificial contrast can occur when a sample consists of reflecting material embedded in an absorbing matrix, for example. SPMs, on the other hand, are largely indifferent to variations in optical or electronic properties, and measure true surface topography.

Despite the apparent simplicity of images from SPMs, SPM images are subject to some artifacts of their own. Fortunately, artifacts are relatively easy to identify.

TIP CONVOLUTION

Most imaging artifacts in an SPM image arise from a phenomenon known as tip convolution or tip imaging. Every data point in an image represents a spatial convolution (in the general sense, not in the sense of Fourier analysis) of the shape of the tip and the shape of the feature imaged. As long as the tip is much sharper than the feature, the true edge profile of the feature is represented.

However, when the feature is sharper than the tip, the image will be dominated by the shape of the tip. Figure demonstrates the origin of tip convolution.

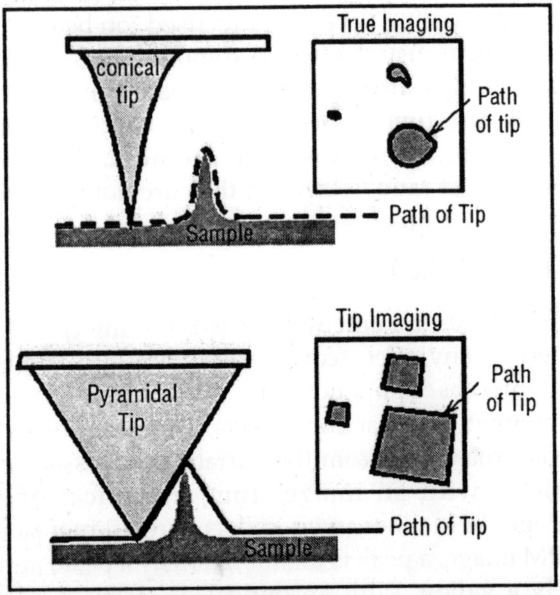

Fig. Comparison between True Imaging and Tip Imaging

The first commercial AFM tips were square pyramidal, formed by CVD deposition of Si3N4 on an etch pit in Si (100). The etch pit is bounded by (111) faces, which means that the resulting tip has a sidewall angle of about 62.5° along the flat side of the pyramid (45° along the corner edge of the pyramid).

If such a tip is used, the edge profiles of SPM images of all features with sides steeper than 62.5° are dominated by the profile of the tip. Tips with higher aspect ratios have been marketed successfully since then, but microfabricated tips remain limited to a sidewall angle of about 80°. Further enhancement of the sidewall angle of the tip can result in degraded durability, or even bending of the tip during scanning.

Because many samples have features with steep sides, tip imaging is a common occurrence in images. Sidewall angles on images should be measured routinely, to determine whether the slope is limited by that of the tip, or truly represents the topography of the sample. One consolation is that the height of the feature is reproduced accurately as long as the tip touches bottom between features. Thus, height measurements and roughness statistics remain fairly accurate. The lateral dimensions from a tip-imaged scan, on the other hand, can provide the user with only a maximum value.

In other words, if we measure a tip-imaged feature as having a width of 200Å, we 'll know that the feature is at most 200Å wide.

To recognize tip imaging, look for a particular shape that is repeated throughout an image. The shape can be different sizes as the tip is convolved with features of different sizes, but it will always maintain the same orientation. If we suspect tip imaging is occurring, we can rotate our sample and image it again. If the tip is dominating the image, the orientation of the tip shape will be the same before and after rotation.

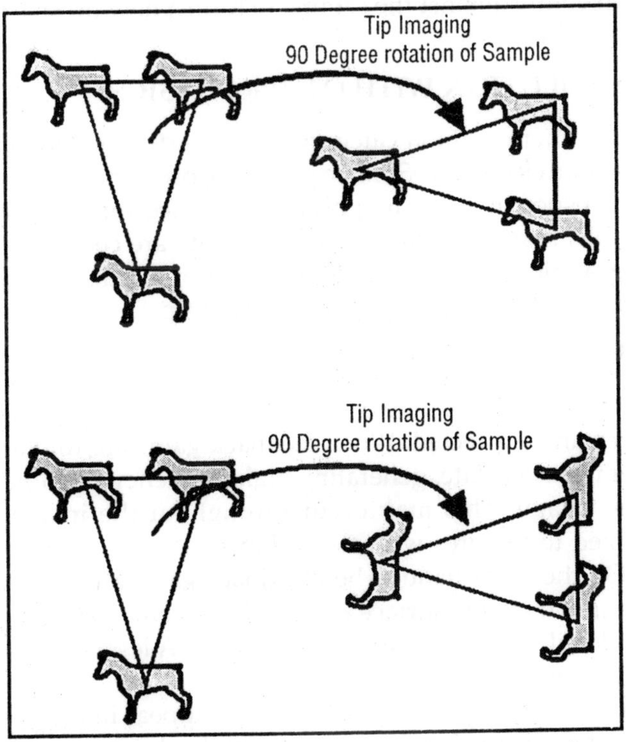

Fig. Top down Image showing Characteristically Shaped Features

If the image is a true representation of the surface, the

shapes in the image will rotate along with the sample. This test is represented schematically in Figure. If our image is dominated by tip-convolution effects, change to another tip.

For STM, the imaging portion of the tip is formed by an atom or cluster of atoms at the end of a long wire. Because the dependence of the tunneling current upon the tip-to-sample distance is exponential, the closest atom on the tip will image the closest atom on the sample.

If two atoms on the tip are equidistant from the surface, all of the features in the image will appear doubled. This is an example of multiple-tip imaging. The best way to alleviate this problem is to apply a voltage pulse to change the tip configuration by field emission. Alternatively, we can press the tip gently against the sample to form a new tip shape, and take another image.

CONVOLUTION WITH OTHER PHYSICS

Another common category of SPM artifacts derives from the sensitivity of an SPM to physical properties other than those responsible for mapping the surface topography. For example, regions of variable conductivity are convolved with topographic features for STM, and soft or elastic surfaces can deform under the pressure of the AFM tip.

FEEDBACK ARTIFACTS

If the feedback loop of an SPM is not optimized, the SPM image can be affected. When feedback gains are too high the system can oscillate, generating high frequency periodic noise in the image. This may occur throughout the image or be localized to features with steep slopes.

On the other hand, when feedback gains are too low, the tip cannot track the surface well. In the extreme case, the image loses detail, appearing smooth or "fuzzy". A less obvious effect is "ghosting".

On sharp slopes, an overshoot can appear in the image as the tip travels up the slope, and an undershoot can appear as the tip travels down the slope. This feedback artifact commonly appears on steep features, represented as bright ridges on the uphill side and/or dark shadows on the downhill side of the feature.

IMAGE PROCESSING CAPABILITIES

Sophisticated image-processing software is available on all commercial SPMs. Beautiful images can be created by introducing artificial light sources, 3-dimensional rendering, colour tables, and curvature enhancement algorithms, by retouching areas of bad data, by filtering environmental noise, and by magnifying or reducing the vertical scale of the image to enhance or minimize surface features of interest. We can flatten an image of a surface of a tiny sphere to look closely at the fine-scale roughness. These are some of the benefits of image-processing software.

Fig. 3-D Rendered Image of a Semiconductor Surface

However, when image processing and image enhancement are used carelessly or irresponsibly, data can be misrepresented.

Fig. AFM Image of Random noise. Left Image is Raw Data

The best way to compare the quality of data sets from

different SPMs is to compare unprocessed "raw" data in a gray-scale image. As an example of misuse of image processing, Figure shows a wholly specious image of a monolayer of an organic film, created entirely from random noise by a carefully crafted filter.

Right image is a false image, produced by applying a narrow band pass filter to the raw data

KEY FEATURES OF SPMS

Considerations outside of the scientific realm are likely to be a part of our choice of an SPM. An instrument that is easy to use will have lower operating costs because a highly skilled operator is not required. Minimal training time is desirable in a multi-user facility. Finally, an instrument that is easy to run will be used more-and produce more results.

USER INTERFACE

Evaluate the complexity of the user interface. It should be laid out in a straightforward, manner with easy-to-use controls. For example, the user interface should have buttons with names that represent commonly performed functions such as "APPROACH" and "IMAGE". After we have seen someone else demonstrate the instrument, ask to operate it ourself. We are then in the best position to evaluate the simplicity of the user interface.

Some SPMs are built upon operating systems that allow multi-tasking. If we can process data while collecting additional data, our throughput will be increased measurably.

OPTICAL MICROSCOPE

All commercial SPMs now include optical microscopes to help monitor the tip-to-sample approach and to select the areas of interest on the sample surface. A good optical microscope speeds up our work because it enables we to position the tip quickly and accurately, exactly where we want to take an SPM image. If we intend to image very rough or oddly shaped samples (for example geological samples, or teeth), or cross sections of any kind that require landing the tip on a narrow edge, a good optical microscope is indispensable. Figure shows a real-time video image of a cantilever positioned over an integrated circuit.

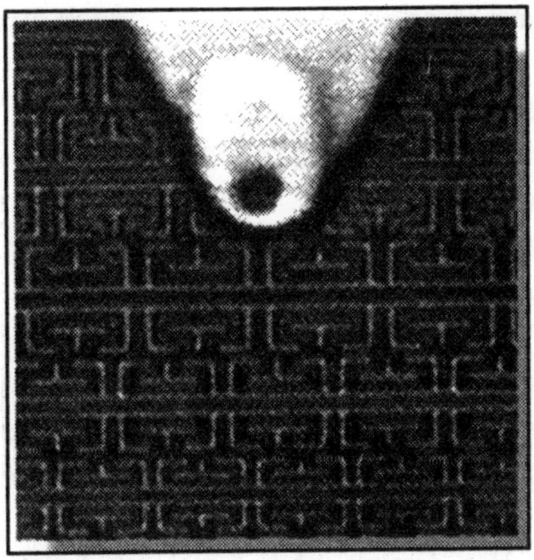

Fig. Real-time video image of a cantilever positioned over an integrated circuit

When evaluating the optical microscope, look for the following features:

- Optical image clarity
- Useful magnification range
- Zoom lens
- Integrated video camera

Motorized focus and zoom controls are timesavers over manual models, but optical image quality is the most important. The magnification should enable we to navigate rough surfaces and cross sections, to a high magnification, where the field of view allows we to place the tip on micron-sized features. Note that a quoted magnification range is subject to "cheating". If we attach a larger monitor to the same video camera, we end up with higher magnification, but no better resolution. The best way to specify the optical microscope is by field of view or resolution.

PROBE HANDLING

Under normal operating conditions, an AFM tip should last for a couple of days, so changing the probe should be a straightforward process. However, the chips upon which cantilevers are mounted are very small and can be unwieldy to handle.

New designs permit pre-aligned, pre-mounted probes to be changed very quickly, with minimal alignment of the beam-bounce detection system.

As with the user interface, we should evaluate how difficult it is to change a probe. Change a probe ourself, instead of watching a highly skilled operator change it.

SYSTEM ACCESSIBILITY

If we intend to perform unique experiments with an SPM, we may be very interested in having access to the software, the electronics, and the mechanics of the system in order to make our own modifications.

Investigate the design of the system to see whether it is open enough for our customized work. Consider also the technical strength and availability of the technical support staff at the factory. We may want their advice more often than a user who is happy to operate the machine off-the-shelf.

Chapter 6

Microtechnology Revolution

Analysts predict that microsystems technology will have a far-reaching influence on device manufacture within the next couple of years. A new generation of injection moulding machines is central to this technology. There's a big future in small things, according to Helmut Detter, head of the department of precision mechanics at the Technical University of Vienna.

As one of the coordinators of the Micro Moulding Machinery project (Wiener Neustadt, Austria), a consortium of companies and organizations developing complete microsystems production solutions, Detter is helping to bring miniaturized device components from the drawing board to full-scale manufacture.

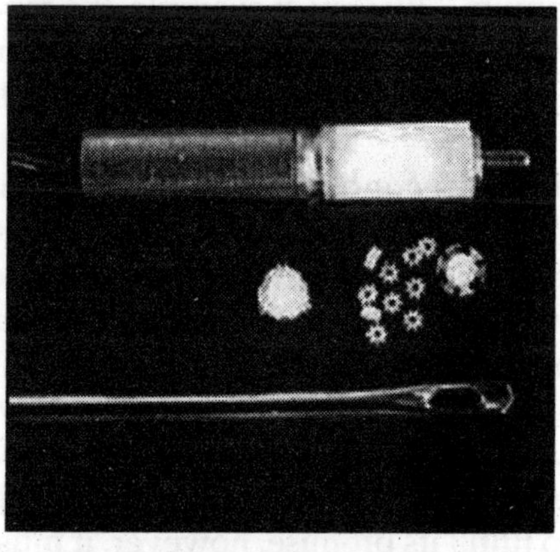

Fig. A Micromotor for Device Applications

One-mm-diam micromotors performing arterial

functions, implantable hearing aids fitted with a 0.002-g sensor, and micropumps for use with implantable drug delivery systems are among the existing and potential applications that he cites.

Detter's confidence in the rapid emergence of microsystems technology (MST) and microelectromechanical systems (MEMS) as driving forces in device design and fabrication is buttressed by a recent market analysis.

A micromotor for device applications was developed and manufactured at the Institut für Microtechnik in Mainz, Germany, in collaboration with Dr. Fritz Faulhaber GmbH & Co. KG. Shown with the component are the individual moulded parts of the motor's epicyclical gear.

The total world market for MEMS and MST is expected to surge from the current US$13 billion to US$38.5 billion by 2002, according to a report commissioned by the Network of Excellence in Multifunctional Microsystems, a body of the European Commission that promotes MST within European industry. While the automotive sector has absorbed most MEMS- and MST-derived products thus far, "the biomedical branch shows the greatest growth and largest potential for further increases," the report states. "It is expected to be the most important segment for MST and MEMS products as we enter the millennium." By way of example, the report's authors predict that drug delivery systems, which currently generate US$10 million in revenue, will become a US$1-billion business by 2002.

The social and economic impact that microsystems technology will have in the next few years could equal and even surpass that of microelectronics, stresses Detter. This is due largely to the breadth of potential applications. "Microsystems technology is multidisciplinary, with micromechanics, microoptics, and microbiology playing an integral role," says Detter. "This creates even greater industrywide potential [than microelectronics]." If the technology is to fulfill its promise, however, it must first bridge the gap from technical feasibility to cost-effective production. Detter again draws a parallel with microelectronics.

"The first integrated circuits were designed at the beginning of the 1960s, but it wasn't until the 1980s that they emerged on a large scale in the marketplace," he says. That was brought about by the development of manufacturing technologies that could produce and assemble these components in an economically viable manner. Similarly, the emergence of microtechnology on a vast scale is dependent on manufacturing systems that can reliably, rapidly, and economically produce the components. For many applications, microinjection moulding has developed into a key enabling technology.

BREAKING THE MOULD

"Today, we have a firm understanding of the problems associated with microinjection moulding," says Holger Moritz, an engineer at the Institut für Microtechnik (IMM) in Mainz, Germany. Established in 1990 to pursue applied research and development and to form partnerships with industry, the IMM's micromoulding department has successfully developed and manufactured products such as micropumps, miniature brackets to hold endoscopic lenses, and flow sensors.

"We initially encountered several challenges in moulding miniature components," says Moritz, the most critical of which involved producing a mould insert. IMM has developed several processes for creating moulds, from milling and electric discharge machining to laser ablation and LIGA technology, which combines x-ray lithography with microelectroforming and micromoulding.

As for the moulding process itself, IMM has resolved such critical issues as positioning accuracy, melt degradation, and a phenomenon known as the "diesel" effect. The latter describes what happens when air, trapped in the mould prior to injection, compresses and heats up. "It can reach 600°C and destroy our polymer," says Moritz.

The diesel effect is neutralized in conventional injection moulding machines by means of a 10-μm-deep slit in the tool. "But a slit of this size would fill up with polymer in our microinjection moulding process, so we needed to develop an [alternative] evacuation system." IMM has also devoted considerable effort to reducing cycle times. "When research into micromoulding started about 15 years ago, the cycle time was roughly 30 minutes," says Moritz. Currently, one to two minutes is standard, he adds. "But I am very confident that we will be down to 30 seconds this year."

A precision bracket designed to hold endoscopic lenses was moulded at the Insutut für Microtechnik in Mainz, Germany.

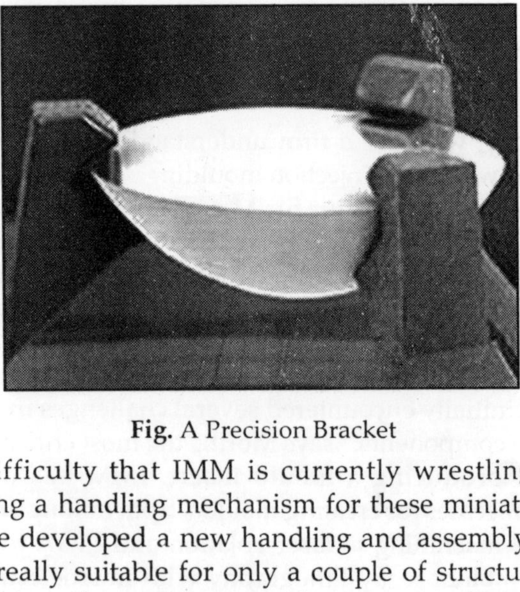

Fig. A Precision Bracket

A difficulty that IMM is currently wrestling with is developing a handling mechanism for these miniature parts. "We have developed a new handling and assembly method, but it is really suitable for only a couple of structures," says Moritz, primarily a micromotor that was developed at IMM in collaboration with Dr. Fritz Faulhaber GmbH & Co. KG (Schönaich, Germany).

Measuring 5.5 mm long and 1.9 mm diam, the motor attains speeds greater than 100,000 rpm. For most applications, a gear is necessary to reduce the motor's rotational speed and to increase the effective torque.

IMM developed a multistage epicyclical gear with an external diameter identical to that of the motor that worked well, but it had to be manufactured in tolerances below 2 μm. This could only be achieved by means of microtechnological fabrication processes, explains Moritz. "Thus far, we have produced about 1500 motors and gearboxes and we plan to install a full production line this year," he says. The project goal is to reach an output of 10,000 units per year. "To achieve that, we must develop moreadvanced assembly automation," says Moritz, "and that is currently the focus of our research."

MACHINE MAKERS START THINKING SMALL

To capitalize on anticipated demand for microinjection moulding machines, some machine builders have developed novel systems designed to consistently, accurately, and rapidly manufacture moulded parts that weigh less than 0.1 g.

Battenfeld (Melnerzhagen, Germany) developed its Microsystem 50, an electrically driven injection moulding machine for miniature parts, within the Micro Moulding Machinery project.

The machine was launched at K 98 in Düsseldorf, Germany, where it was shown moulding a polycarbonate part weighing 0.002 g that is designed to house a sensor and to be implanted in a human ear.

According to Martin Ganz, manager of the company's microtechnology division, other potential medical applications for the system include micropumps and minimally invasive surgical instruments.

Because of process parameters that are unique to micromoulding, Battenfeld chose to develop a new system rather than modify an existing machine.

Conventional equipment uses large sprues and runner volume to minimize material dwell times and prevent melt degradation, according to Ganz. Consequently, the required heating and cooling stages lead to higher energy consumption and longer cycle times. In addition, long flow paths in relation to the part size require the use of more-powerful injection units. Battenfeld circumvented these limitations by incorporating a 5-mm-diam plunger injection unit and a rotary table into the system.

This POM microsensor housing weighing 0.0028 g was moulded on Battenfeld's Microsystem 50.

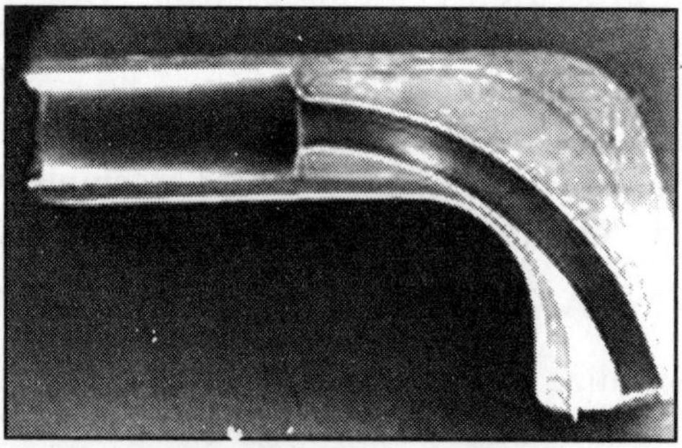

Fig. POM Microsensor

"The injection device, which works like a plunger, reaches the splitting line of the mould," says Ganz. This design enables injection of the material in close proximity to the cavity, resulting in a very small sprue. Because there is less volume to cool down, shorter cooling times are achieved. In addition, a rotary table system enables cavity filling to take place at one station while the parts are ejected at the other. According to Ganz, these innovations have helped to reduce cycle times for the production of microparts by as much as 50 per cent compared to conventional systems.

The parts are removed by means of a two-axis system fitted with suction cups that also orient the parts for quality control and packaging operations. The entire system is contained within a cleanroom enclosure. Dr. Boy GmbH (Neustadt, Germany), which introduced a 12-t injection moulder with a 0.3-g shot weight in 1998, has announced the launch of a new model that can produce parts weighing as little as 0.0015 g.

"The Boy 12 A micro will mould a minimum shot weight of 0.15 g," says the company's UK sales manager Ian Crawford. Because the moulds can contain as many as 100 cavities, particular attention has been paid to quality control processes. The machine includes a CCD camera with a vision system to monitor cavity filling and parts-ejection processes along with other quality control parameters.

The primary feature of the Boy 12 A micro, according to Crawford, is the screw plasticizing unit with needle injection.

"The needle in the screw ensures that even the smallest shot weights are accurately metered, and it prevents problems associated with bypasses and dead corners that may arise with downstream plunger injection," he says.

The relatively large screw ensures optimal homogenization and constant plasticizing time; the screw geometry eliminates feed problems that may be encountered with the use of smaller screws. In addition, extended dwell time, which can adversely affect heat-sensitive materials, is reduced by use of the first-in, first-out (FIFO) principle.

Ferromatik Milacron S.A. (Maurepas, France) has designed a microinjection unit that can be adapted to fit the company's K-series machines with clamping pressures from 40 to 110 t. "The points we wanted to address in developing this unit were moulding accuracy, feed problems, and dwell times in relation to thermally sensitive materials," explains sales engineer Jean-Pierre Petel.

Fitting the unit with a specially designed 18-mm screw solved many of these problems, according to Petel. "This screw diameter lets us use standard-sized pellets, but it does pose a problem regarding dwell time."

Fig. Microsystem 50, Designed to Mould Parts Weighing Less than 0.1 g

Some companies have dealt with this by using a 14-mm screw, says Petel, "but then we have feeding problems because the screw is too narrow to process standard pellets." Ferromatik Milacron chose to reduce the screw's length/diameter ratio from the customary 20 or 22 to 15. "Consequently, the material will not stagnate in the screw and we reduce the risk of thermal degradation," says Petel. The unit's needle piston works according to the FIFO principle to further prevent extended dwell times.

Battenfeld recently introduced its Microsystem 50, which is designed to mould parts weighing less than 0.1 g.

The microinjection unit is currently being used at IMM, Petel adds, for the production of moulded medical and electronic parts.

Microstructures are destined to play an enormous role in medical technology, says Detter, and microinjection moulding is clearly the process of choice for the reliable and cost-effective fabrication of many of these plastic components. While purpose-built machines have eliminated many of the difficulties involved in moulding micron-sized parts, obstacles remain in the automation of parts handling and assembly operations. New materials adapted to the demands of microinjection moulding and the intended use of the final product will also become a key issue as MEMS- and MST-derived devices enter the mainstream in the new millennium.

MICROMATERIALS

"Microtechnology uses micro amounts of materials," says Holger Moritz of the Institut für Microtechnik in Mainz, Germany, and that is developing into a problem for companies moulding miniature components. "We are unable to find suppliers who are working on materials with properties suited to this technology," says Moritz. The cause is primarily a matter of economics. "Polymer suppliers think in hundreds of tonnes, but it only takes one tonne of material to produce about 100 million of the gearboxes we are currently making for a micromotor," explains Moritz.

Engineers have learned to work with commodity resins, stresses Moritz. "We have successfully processed about 40 different types of polymers so far at the institute. We can build microstructures in PEEK, polycarbonate, POM, to name a few," he says, adding that the best results are achieved with low-viscosity grades. PTFE, however, has proven resistant to micromoulding and Moritz regrets that compounds are not suitable.

"The compound component is as big as some of the parts we mould, so we see the problem," says Moritz. "It's unfortunate that we cannot yet use reinforced materials with enhanced properties in some of these projects." IMM is participating in research to develop nanofilled polymers, but the endeavour is faced with a formidable obstacle: the tiny asbestoslike fibres may be carcinogenic. "It's not because of the material per se, but because of the fibres' tiny dimensions," says Moritz.

Nevertheless, future breakthroughs in microtechnology will rely partly on developments in materials science, according to Helmut Detter, head of the department of precision mechanics at the Technical University of Vienna. "Mass-produced polymers will not be sufficient for the fabrication of parts in a dimension of micrometers," says Detter. "We will need a new type of high-value, small-quantity material," he says, adding that his department is currently investigating a material used in the manufacture of stents that has promising properties for micromoulding applications.

US FDA'S QUALITY SYSTEM REGULATION

The MDD's full quality assurance route to CE marking may provide manufacturers with a shortcut to obtaining US FDA approval. European medical device companies that have met the requirements of the Medical Devices Directive (MDD; 93/42/EEC) and wish to market their products in the United States are faced with the challenge of satisfying US FDA's new regimen—Current Good Manufacturing Practices (CGMP) and quality system regulation (QSR). The stakes are high. With 260 million consumers, the US market for medical devices and diagnostic products is projected to reach $64 billion in 1999. The critical question for many device manufacturers is how to bridge the gaps between the two regulatory systems.

There is no simple, single answer. Although the new regimes are part of a trend toward harmonization of regulatory requirements designed to eliminate trade barriers, the two systems are not perfectly aligned. One manufacturer with a newly minted CE mark may be able to achieve FDA compliance relatively easily; another conceivably might have to start over from scratch.

Such anomalies are essentially the result of two primary misalignments between the systems. Both the European Union and the United States divide medical devices into different classes and provide for somewhat different requirements for each class. As a result, closing the gaps between US and EU requirements can vary by product classification.

Fig. Medical Devices into Different Classes

US FDA's QSR offers one route to quality assurance for a given medical device class. By contrast, the European Union has adopted a modular approach to conformity assessment for quality assurance, with up to four different routes to quality assurance for a Class IIb device, three for a Class III product. A European manufacturer that follows the EU route that most closely parallels US FDA guidelines could save considerable time meeting regulatory requirements.

A Tale of Two Systems

Underlying the complexity and differences between the two regimes are some fundamental similarities. "Both systems have the same two goals," explains Ron Belmont, director of food, drug, and healthcare support activities for Excel Partnership Inc., an international training and consulting firm based in Sandy Hook, CT, USA, with European headquarters in Nantwich, Cheshire, UK. "Their objective is to ensure that a medical device company produces a safe product and that it is able to provide quality assurance that it can manufacture this safe product consistently," says Belmont.

Both the European Union and the United States require the development of sufficient technical documentation to enable regulators to determine whether the product or product family as designed and conceived is safe.

The closest US FDA equivalents to the technical file for Class I, Class IIa and IIb products, and the design dossier for Class III devices is the premarket notification 510(k) evaluation and premarket approval (PMA) review. The former covers established devices and products that are largely similar to devices already on the market; the latter is generally required for Class III and high-risk Class II devices. For US companies it can be very difficult to close the gap between the 510(k) and the technical file. For European manufacturers, however, the technical file should more than satisfy US FDA in most cases.

Kolka points out that US FDA addresses many of the same questions as do the essential requirements in its recently published *Final Design Control Inspectional Strategy*.

"Any manufacturer—whether American, European, or Japanese, for that matter—that is developing new devices for international markets is well advised to follow the essential requirements in creating technical documentation," he says. "By meeting CE-marking requirements, we will also largely satisfy US FDA, which readily accepts EU harmonized standards and European national standards to demonstrate compliance with its requirements. By following this course of action, we satisfy both regulatory regimes in one fell swoop," says Kolka.

The quality assurance system in both the European Union and the United States must cover both production and design control for Class II (both IIa and IIb) and Class III products. According to US FDA's QSR, this requirement can only be met by ISO 9001, the international quality system standard for both design control and manufacturing, and ISO 13486, which applies ISO 9001 to the medical device industry. The European Union has also adopted the same standards (designated as EN ISO 9001 and EN 46001) as a route to meet conformity assessment requirements. The option, called the full quality assurance system, is described in Annex II of the MDD.

However, the European Union also offers other options to manufacturers of Class II and Class III devices as part of its modular approach to conformity assessment for quality assurance. These options, which are described in Annexes III, IV, V, and VI of the MDD, were developed to give companies separate design control from manufacturing or production control. In some cases, European manufacturers might benefit from the flexibility inherent in these alternatives; unfortunately, none are acceptable to US FDA.

The Shortest Route to Quality Assurance

"Any European company that plans to do business in the United States should elect the full quality assurance system option when going for CE marking," explains Belmont. "Any other alternative is self-defeating, since the manufacturer would then have to meet the requirements of two systems."

The plot thickens somewhat for Class I devices. US FDA exempts most Class I devices from its design control requirements and requires manufacturers to satisfy the QSR by meeting ISO 9002 and ISO 13488 (EN 46002 in the European Union), the international standards that apply to production systems and exclude design control.

As with Class II and Class III devices, the situation with the MDD is more complex. For Class I sterile and measuring devices, the European Union offers several options. Only one of these overlaps with US FDA. It combines EN ISO 9002/EN ISO 13488 (Annex V, production quality assurance) with a self-declaration by the manufacturer that the product conforms to the design specifications in its technical file (Annex VII). This is the most reasonable option for EU manufacturers of Class I sterile and measuring devices seeking to sell their products in the United States. Class I nonsterile, nonmeasuring devices are a separate case. The MDD accepts self-declaration (Annex VII), but US FDA requires compliance with ISO 9002 and ISO 13488. Thus, an EU manufacturer of CE-marked nonsterile and nonmeasuring devices would not automatically satisfy US FDA requirements.

"Although some European medical device manufacturers may have hurdles to overcome in introducing previously CE-marked products into the United States, manufacturers developing new devices can simultaneously meet the demands of both the European Union and US FDA," says Belmont. "So, in that sense, the two regimens are moving toward harmony, and trade barriers are coming down."

EU RECYCLING LAWS

EU recycling laws may cost industry at large billions of dollars per year in fees, data collection, and tracking, according to experts participating in a workshop on the price of compliance with European packaging mandates. Larger multinational companies will spend an average of US$500,000 annually to set up a package tracking system and another US$100,000 to maintain it, according to UK-based consultant Michael Coe who participated in the workshop.

Under take-back laws implemented in the European Union and a dozen other countries worldwide, manufacturers must join a collection organization (and pay concomitant fees) or retrieve their packaging. According to some workshop participants, compliance with the EU's packaging legislation, which will take effect in 2000, can be as high as US$10 billion per year. Preparing the documentation, according to the president of Environmental Packaging International Victor Bell, frequently is more onerous than the actual fee paid to the collection organization. Fees are paid to industry recovery organizations in Europe and Asia, but the definitions of packaging and composites, along with the dollar amounts vary wildly, he adds.

One thousand folding cartons with less than 5 per cent coating, said Bell, would be assessed US$22.90 in Germany,

US$1 in Belgium, and US$5.10 in France. A foil/plastic laminate would cost US$120 in Germany, US$40 in Belgium, but the same US$5.10 in France.

EU packaging legislation will require a written assessment that the package has been source reduced, that it is recyclable, reusable, or recoverable, and that it will not have a negative impact on the countries' recovery systems.

SOFTWARE SYSTEM FACILITATES MEDICAL-IMAGE TRANSFERS

Etiam (Rennes, France) has developed a software system for cooperative medical imaging. The system, called Mediem (medical imaging by e-mail), can be integrated as a component in imaging equipment and software.

Mediem allows for multimedia exchanges between healthcare professionals by enabling standard medical images to be transferred via the Internet. Current users of Mediem include healthcare companies, hospitals, and manufacturers of medical-imaging equipment and software. Although Mediem currently is not sold to physicians, the company is exploring this market. "The company is in a constant state of research and development," says Etiam sales engineer Florence Michelon. "We want our solution to become the standard."

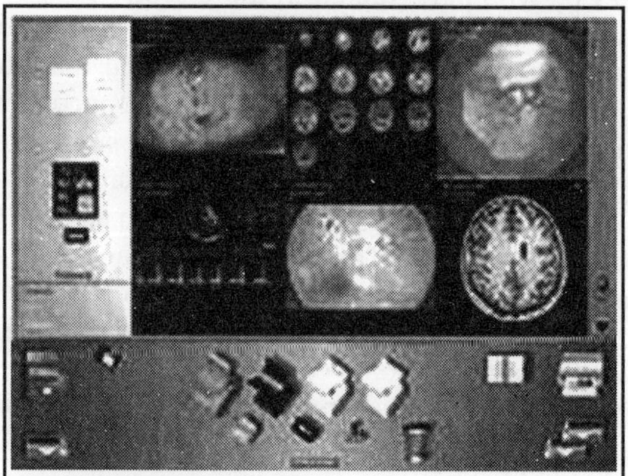

Fig. Medical Image Transfer Software

At the system's core is its ability to send, query, retrieve, and view digital imaging and communications in medicine (DICOM) images of any form via e-mail. The DICOM format, a standard in telemedicine, begins all files with a header that includes an identification of the patient and the physician performing the diagnosis.

Mediem can access DICOM images—from medical equipment, the Web, or a local network—and retrieve them for storage or modification. The files can be annotated with text, video, or audio commentary and accompanied by medical reports. Another significant feature is the software's incorporation of encryption techniques that permit secure, confidential transfers of medical images.

The software was developed in Java, a platform-independent Internet programming language, and supports several standard file formats and transfer protocols, including HTML, XML, RTF, HTTP, SMTP, POP3, and S-MIME. The DICOM medical-imaging standard is supported in all modalities, including black and white, compressed, and single-frame or multiframe.

Etiam was founded by Cerium, a group of engineers that have been responsible for software development in cooperation with the Signals and Medical Imaging (SIM) laboratory and hospital of the University of Rennes as well as industrial companies.

Etiam strives to provide high-level software libraries to firms that are developing medical-imaging products. The company's clients include GE Medical Systems (Buc, France) and France Telecom. Etiam presented Mediem for the first time at the RSNA 1997 show hosted by the Radiological Society of North America in Chicago. Future versions of the system are in development and will include teleconference features.

The company anticipates the widespread use of Internet technology for medical imaging in the coming years and hopes that its software will play a key role in the development of this market. "We are really at a crossroads," says Michelon. "Etiam combines expertise in software engineering, communications, and medical imaging, so we are in a position to work on establishing standards for the integration of these three sectors."

NOVEL NEURODIAGNOSTIC INSTRUMENT

A full-service contract manufacturer of medical devices, Nova Biomedical Corp. (Waltham, MA, USA) announced that it has reached an agreement with NeuroMetrix Inc. (Cambridge, MA, USA) to manufacture, package, and distribute a neurodiagnostic instrument currently under development.

The NC-stat monitor will allow physicians to accurately, rapidly, and cost-effectively measure nerve-conduction parameters used to diagnose and evaluate systemic and entrapment neuropathies. A low-cost handheld microprocessor-based device, the system consists of single-use disposable biosensors and an electronic monitor. NeuroMetrix, a spin-off from the Harvard-MIT Div. of Health Sciences and Technology, announced that shipment of the product will begin early this year.

Fig. Novel Neurodiagnostic Instrument

In addition to offering development, design for manufacture, production, distribution, and repair services to medical device OEMs, Nova Biomedical produces clinical chemistry analyzers for critical-care applications using its proprietary ISE sensors. These sensors are also used in OEM products that the company routinely manufactures for other diagnostic companies.

ISMECA MEDICAL DEVICE ASSEMBLY UNIT

Automating assembly processes can reduce production costs while boosting quality assurance. What is the biggest challenge facing potential suppliers to medical device OEMs? For Ismeca Europe S.A., a company that manufactures automated assembly systems, neither complying with industry's stiff regulatory requirements nor engineering systems that can assemble delicate miniature components present a serious obstacle. "We have acquired the know-how to deal with those issues," says CEO Bernard Lacoste.

The tough part, according to him, is acquiring name recognition and gaining industry's confidence. To show its commitment to the device industry and to earn its trust, Ismeca has established a global business unit devoted to medical device assembly.

The youngest of the company's three divisions—the other two are focused on back-end equipment for the semiconductor industry and assembly automation—the medical device unit is also the firm's fastest growing one. Within three years, Lacoste anticipates that it will generate approximately 25 per cent of the company's revenue. The business plan is ambitious, he concedes, but realistic. His optimism is rooted in what he terms an unusually good match of industry's needs and Ismeca's capabilities.

Faster Processes, Smaller Assemblies

Initially a supplier of automation equipment to the watch-making industry, the 35-year-old company began producing back-end equipment for semiconductor applications in the 1970s. "From there we went into assembly automation, with an emphasis on small or delicate parts, and, increasingly, the assembly and processing of plastic parts," says Lacoste.

"On the one hand we were pushed by the semiconductor industry to accelerate processes while increasing quality and reliability," he says. "But we have also demonstrated our expertise in designing systems that perform precision assembly operations of small components." In other words, adds Lacoste, a tailor-made résumé for a supplier to the device industry.

One feature of the company's assembly automation systems with great potential appeal to device manufacturers is the modular architecture. "The Smart flex concept lets manufacturers start small in size, say 2 X 1 m in floor space," says marketing and product manager for Medical Assembly Dirk Dauw. "As their needs evolve, they can easily add on to that initial piece of equipment." The average amortization time for Ismeca's automation equipment, adds Dauw, is two to three years. Under pressure from purchasers, device manufacturers are being forced to find new ways to reduce production costs, he adds. "The only way to bring down those expenses is through automation," he stresses.

The other benefit of automating processes, Dauw adds, is the immediate gain in quality. "Manual assembly will never achieve the level of quality assurance that is required in the manufacture of certain medical devices," says Dauw.

Fig. Smart Flex

Ismeca's systems also feature high reliability, according to Dauw, which is partly a result of the use of cams and servos in lieu of hydraulic components. The superiority of electronic systems over mechanical ones is borne out by Ismeca's track record in the semiconductor industry, "where reliability is just as critical as it is in device manufacturing," says Dauw.

To illustrate the company's expertise in devising automated solutions for complex assembly procedures, Lacoste described some of the company's recent projects. One application involved feeding, orienting, inserting, and bonding a 15-mm-long cannula to a hub.

"The cannulae, which weighed very little and had a diameter of 0.25 mm, were difficult to process. So we had to develop novel approaches for most of the operations," he says. Compounding the difficulty, the tip had to be set at a precise distance from the hub. "We developed a vision-guided insertion method that achieved assembly tolerances of less than 0.025 mm."

Another medical application entailed the ultrasonic welding of a diagnostic cartridge to extremely stringent requirements on a high-speed pallet-based line. "The long weld path had to be hermetically sealed, and the melt rate and weld parameters had to be near perfect," says Lacoste.

Ismeca determined that a weld of this quality could not be performed if the product remained on the pallet. "We built a rigid single-weld anvil," explains Lacoste. "The parts were transferred from the pallet onto the weld anvil, and with each cycle a previously welded assembly was returned to the pallet." The system, he adds, achieved essentially perfect welds at full line speed.

Flexibility and Speed

The company has also designed systems suited for the processing of a family of products. "We frequently run into situations where the production volume of one single configuration is not sufficient to justify automation," says Lacoste. Ismeca circumvents this problem by using electronic lockout systems that restrict access to station areas or side loops that route pallets along a "detour." Both of these solutions feature almost instantaneous changeovers. The company can also supply replacement stations, in which case changeover times can vary from 10 minutes to one hour.

In addition to Switzerland, Ismeca has manufacturing facilities in the United States and Malaysia, and sales offices in Switzerland, France, Italy, Germany, the United Kingdom, the United States, and Hong Kong.

REXAM MEDICAL PACKAGING EXPANDS CAPABILITIES

As part of its ongoing investment programme, Rexam Medical Packaging Ltd. recently added an air-knife coater and coating and lamination pilot line to its manufacturing facility.

The wide-width coater will enable the company to manufacture peelable coated Tyvek and paper products that were until now produced at the company's US facilities for worldwide distribution.

The development and manufacture of new products will also be a priority, according to market manager Roger James. "Our development programme extending the range of air-knife coated products is nearing completion," he says, "and many of them will provide potential cost-reduction options." Air-knife coated products are typically used for packing large and heavy products or those with sharp edges such as dialyzer sets, cannulae, and syringes.

The new coating and lamination pilot line is designed to process Aclar products. This will enable the company to offer low-permeability moisture barriers from 0.12 to 0.34 $cm^3/m^2/$day using a range of barrier Aclar products. These products will complement Rexam's existing range of PVDC-coated polyvinyl chloride barrier blister-packaging materials. The company offers a variety of PVC products that meet differing contact and approval requirements.

IEC VIDEO EXPLAINS FAST-TRACK SPECIFICATIONS

Industrialists and the president-elect and general secretary of the International Electrotechnical Commission (IEC) team up to explain the rationale behind IEC's high-speed specification process in a recently released video. Industry Technical Agreements (ITAs) are designed to meet an urgent market need for standards principally in fast-moving technology sectors where products tend to have short life cycles. Large companies, a consortium, or indeed any group of interested parties are eligible to create an ITA, according to the IEC. Once a consensus has been reached by the group, the agreement becomes a de facto specification.

ITAs potentially apply to all domains of IEC work, including electrical equipment intended for medical use, says IEC information manager Dennis Brougham. "It should be stressed that ITAs are primarily intended for areas where the traditional standardization process occasionally has difficulty keeping pace," he adds. ITAs can be drawn up in a matter of months rather than years. "However, if someone suggested an ITA for electrical equipment in the medical domain, I feel that we would be receptive to the idea," says Brougham.

The first ITA, which is currently under way, is aimed at defining specifications for standardized multimedia platforms. When completed, it will allow end-users to access a range of multimedia services from a single platform. The Open Platform Initiative for Multimedia Access comprises more than 40 companies and organizations and is scheduled to be implemented by September 1999.

CZECH GROUP OFFERS OUTSOURCING ASSISTANCE

If we are an OEM looking for a supplier of contract services, the Czech Subcontracting and Partnership Exchange (S-Centrum; Prague) is eager to help. The organization will source suitable national partners for a nominal fee, according to Vlastimil Vejnar of the Confederation of Industry of the Czech Republic. The programme is supported by the European Union and the United Nations Industrial Development Organization.

"Companies interested in using this service simply need to provide us with information on the type of services they are looking for and the expertise that is required, and we will locate an appropriate enterprise," says Vejnar. S-Centrum has conducted comprehensive audits of its member companies and established a database to optimize the selection process. Information related to quality systems, manufacturing capabilities, capital equipment, and other data has been captured and coded in accordance with EU nomenclatures. "Consequently, we can select companies based on a combination of clearly defined criteria," says Vejnar.

S-Centrum can source suppliers involved in metal fabrication, plastic products, electronics, and textiles, among other sectors.

ANALYTICAL INSTRUMENTS

Application-Specific Meter Performs Tests on Multiple Analytes

An application-specific meter capable of performing tests on multiple analytes can be programmed to accommodate varying reagent batches and features. The PART (Personal Analyzer Rapid Test), developed by LRE Technology Partner GmbH (München, Germany) for OEM use, is designed for both wet and dry chemistry testing.

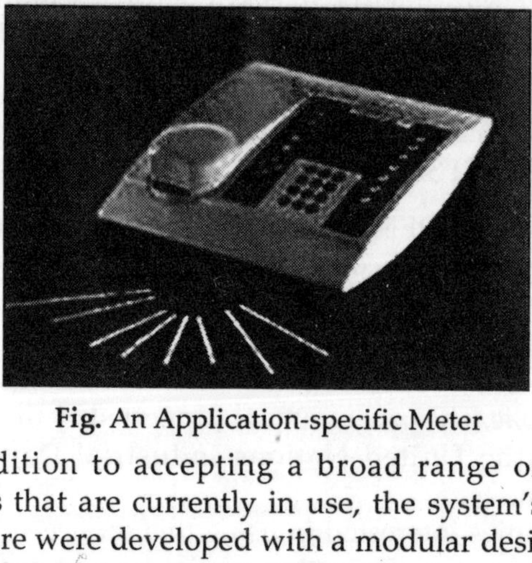

Fig. An Application-specific Meter

In addition to accepting a broad range of test-strip parameters that are currently in use, the system's hardware and software were developed with a modular design to make it compatible with tests that are still in development. Although the device has numerous applications in the medical, environmental, and industrial sectors, one especially promising area is in point-of-care diagnostics, according to product manager Joachim Dyck.

"Point-of-care diagnostics is the wave of the future," says Dyck. Patients increasingly will demand quantitative screenings in their doctor's office, he predicts, and a device such as the PART can help to make this feasible.

"The software gives quantitative results, not just pass or fail, and this allows the practitioner to follow the patient's progress," says Dyck. Another salient feature for this type of application, he adds, is PART's modular design. "A gynecologist and a urologist will not be running the same tests, and this user-programmable analyzer can accommodate their varying needs."

The desktop analyzer measures 262.7 X 244.8 mm and is only 72.4 mm high. The system can be configured for electrochemical and transmissive measurements, and evaluation software for strip calibration and for quality control of strip production—one of the analyzer's core applications—is available.

The company recently expanded its product range to include high-end optical systems through a partnership with J & M Analytische Mess- und Regeltechnik GmbH (Aalen, Germany). J & M, which makes a spectrometer that measures cuvettes without light loss, develops and produces diode-array spectrometers for a range of applications.

INJECTION MOULDING

Hot-Runner Systems Reduce Material Waste, Increase Throughput

To precisely meet an application's processing criteria, mould size, and design as well as varying parts geometry, a company supplies custom hot-runner systems in service voltages of 230, 24, and 5 V.

More than 1000 different nozzle types are available from Günther Heisskanaltechnik GmbH (Frankenberg, Germany), and intercavity pitches down to 11 mm are possible. According to marketing manager Eckart Spork, this capability has contributed to the company's growth as a tooling supplier to the device industry.

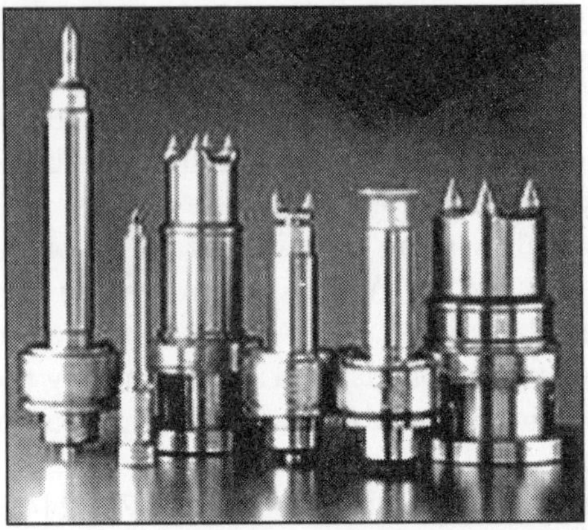

Fig. Injection Moulding

In one recent project cited by Spork that involved the manufacture of an eyedropper part, "Günther's hot-runner injection moulding process resulted in a 50 per cent productivity gain, compared to cold-channel technology." Another advantage of hot-runner systems is the reduction of material waste, he adds. "This is especially significant in heavily regulated industries, like medical, where lost material may not be recycled," says Spork.

The company has also succeeded in eliminating some of the traditional drawbacks of hot-runner systems, such as restrictions on the types of materials that may be processed, involved maintenance procedures, and susceptibility to wear.

All of Günther's hot runners are externally heated and can be operated using the same controller. Building upon this architecture, the company recently introduced a modified 230-V system that can process thermally sensitive polymers. New heating elements and revisions in the nozzle adjustment area ensure that a homogeneous temperature is maintained as the material flows through the system.

Modular nozzles allow most maintenance procedures to be performed on-site. And to reduce wear, Günther uses hard metal tips on the nozzles. The material's heat-conductive properties are specifically suited to thermally sensitive plastics that require high processing temperatures, according to Spork.

The 230-V manifold systems can be configured in h, cross, star, or t shapes, and can be fitted with nozzles of any service voltage.

MATERIALS

Liquid-Crystal Polymer Rivals Stainless Steel for Tubing Applications

Liquid-crystal polymer (LCP) tubing can be made three times thinner than tubing formed of polycarbonate yet still match the stiffness of the latter. LCP tubing, which typically offers twice the crush resistance of conventional plastic tubing, offers a potential alternative to stainless steel in such medical devices as endoscopic and laparoscopic instruments and biopsy needles, according to Precision Extrusion Inc.

Fig. Liquid-Crystal Polymer Rivals Stainless Steel for Tubing

To produce the tubing, the company employs a proprietary rotating-extrusion-die technology that was developed and patented by Superex Polymer. "This process," says Precision Extrusion president Mike Badera, "puts a particular orientation on the molecular chains in the polymer."

It is this biaxial orientation, he says, that enables his company to modify such properties as wall thickness and tensile strength and to overcome certain structural limitations of conventionally extruded LCP. Both the grade of LCP used and the wall thickness determine the tubing's strength range.

"Because of its unique properties," says Badera, "LCP is being used more and more in the injection moulding process for the production of complex parts." The USP Class VI–approved material withstands repeated sterilization via autoclave, chemical, EtO, or radiation methods. Its additional benefits include dielectric strength of 1100 V/mil, a high melt temperature in excess of 500°F, use temperatures greater than 400°F, more than 20,000 psi tensile strength, and a flexural modulus of 2 million psi.

According to Badera, a patent is pending for a secondary operation—postextrusion polymerization—designed to further strengthen and improve the LCP material. Amalgamations of LCP and such materials as polycarbonate, he adds, are currently undergoing testing to determine their suitability for use in a variety of applications including catheters.

SPOTLIGHT ON FILTERS

IV Filter

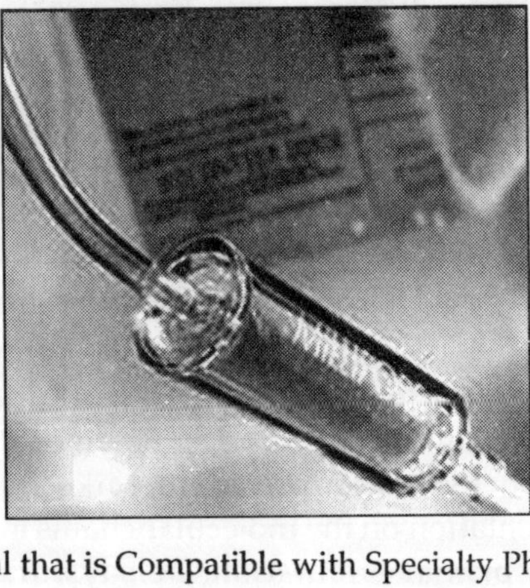

Fig. A Material that is Compatible with Specialty Pharmaceuticals

Composed of a material that is compatible with specialty

pharmaceuticals, an in-line IV filter can be used for up to 96 hours for the retention of particulates and microorganisms. The IV Express is a vented 0.22-mm filter that is suited for use with antibiotic solutions or in situations where the proper alignment of particulates and multiple admixtures is critical. The company specializes in the development of small custom filtration units for reproducible low-volume applications in OEM devices.

In addition to liquid- and gas-filtration products, the firm supplies Quick Release media for diagnostic assay kits. Millipore S.A., Saint-Quentin-Yvelines, France.

Woven Wire Mesh Filters

Fig. Woven wire Mesh Filters

Woven wire mesh filters, screens, and strainers can be fabricated in a variety of materials to almost any shape or size. They are supplied in small, medium, or large production runs. Suited for engineering, processing, and OEM applications, the components can be produced to customer specifications. Design and development assistance is also available. In addition to woven wire mesh, perforated plate and other combinations can be produced. Endecotts Ltd., London, UK.

Medical venting laminates

A company provides ePTFE laminates for medical venting applications. The Gore-Tex laminates can be incorporated into transducer protectors, IV filters and spike vents, urine and ostomy bags, respiratory filters, and suction canisters.

Fig. Medical Venting Laminates

The company assists in determining the appropriate laminate for specific applications. W.L. Gore and Associates GmbH, Putzbrunn, Germany.

Microfilter

Fig. Microfilter

Combining high airflow and low resistance, a microfilter is suited for a variety of low-pressure air and low- to high-vacuum applications.

Filter media ranging in size from 500 to ¾0.1 μm can be used. Hydrophobic properties allow free passage of air while stopping fluids. A clear plastic housing enables users to verify filter integrity and the presence of contaminants. The company also supplies a miniature gas-line filter that features the same specifications in a smaller package. Performance Systematix Inc., Caledonia, MI, USA.

Heat-Moisture Exchange Filters

Fig. Heat-moisture Exchange Filters

Recently developed heat-moisture exchange filters achieve 34.6-mg moisture output, reportedly one of the highest rates ever recorded.

The company manufactures a range of filters that attain bacterial efficiencies of 99.99999 per cent for use in oxygen concentrators, anaesthesia and ventilation equipment, nebulizers, autoclaves, sterilizers, and other devices. Pleated filters are subjected to a sodium flame test to determine efficiency. High-capacity hydrophobic filters in disposable and autoclavable configurations and other special requests can also be accommodated. Air Safety Ltd., London, UK.

Pipette Filters

Pipette tip filters manufactured using a fibre-conversion process provide protection from cross-contamination of samples caused by aerosol or radioactive particle transmissions. Depth-type filtration facilitates the coalescence of aerosol particles into larger droplets that are more easily captured by the Transorb filters.

The products are designed and manufactured to meet customer requirements related to physical dimensions, filtration capture rates, flow resistance, and other performance specifications. Filtrona Richmond Inc., Richmond, VA, USA.

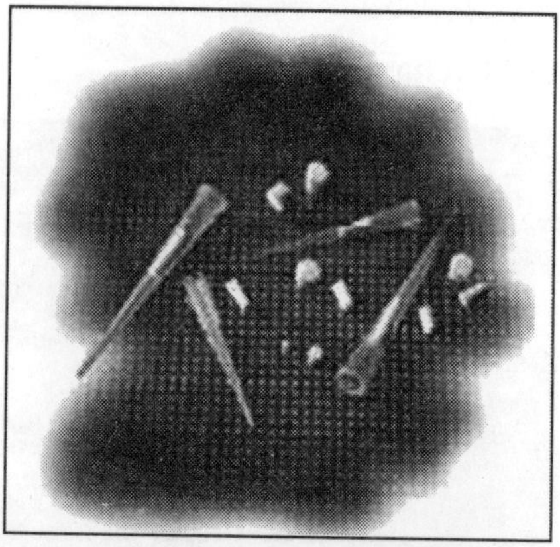

Fig. Pipette Filters

Pyrogen-Free Materials

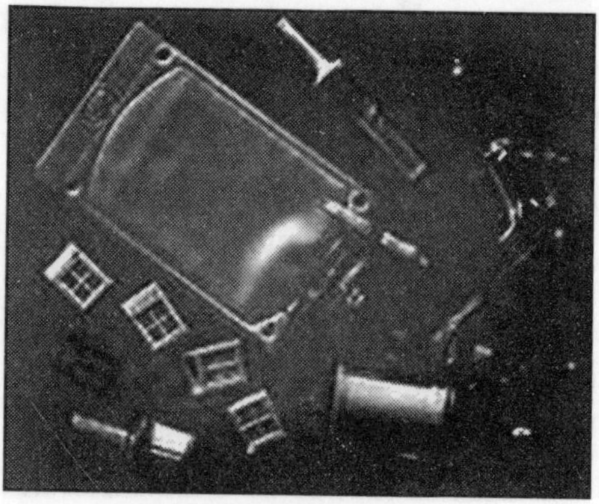

Fig. Pyrogen-free Materials

Fabrics and components suited for medical filter and diagnostic applications are manufactured according to CGMP guidelines in ISO 9000–certified facilities. No water is used to clean these pyrogen-free products. Mesh components are fabricated in a Class 10,000 cleanroom. Plasma treatment and other surface modification processes are available. SaatiTech S.p.A., Veniano (CO), Italy.

Subminiature in-line Filters

Fig. Subminiature in-line Filters

Disposable subminiature in-line filters are designed to remove most particulate contaminants in compressed-air systems. The filtration medium—a Dutch-weave stainless-steel screen—is available in 5, 25, 43, and 73-μm sizes. Three filter models as well as filtered orifices in sizes ranging from 0.004 to 0.30 in. can be supplied. Air Logic, Racine, WI, USA.

INNOVATIVE NEW PRODUCTS

LABELS

Multipage Labels help Obtain Compliance with Regulatory Requirements

Machine-folded multipage leaflets combined with a self-adhesive label are supplied on-reel for use with standard labelling equipment. While Fix-a-Form labels, licensed worldwide by Fix-a-Form International (Bury St. Edmunds, Suffolk, UK) are not new, their popularity has surged within the pharmaceutical, device, and diagnostic industry as legislation requiring more-detailed product information in several European languages is implemented. Fix-a-Form labels were exhibited at Emballage by the French licensee, Pago S.A. (Trappes, France), which also offers custom labelling solutions and a variety of labelling equipment.

According to Fix-a-Form marketing manager Stephen Jarrold, the leaflet-style labels have achieved tremendous market share in the pharmaceutical sector in the past two years. "Growth has been between 20 and 40 per cent," he says, adding that there is much more to come. "Many of our licensees are creating special production areas and cleanroom environments exclusively for the production of pharmaceutical labels. Leaflet-labels are now fully accepted in the pharmaceutical sector because they are recognized internationally," says Jarrold.

In addition to providing space for adequate product description and diagrams and pictures where necessary, the leaflet-style labels can be overprinted with batch numbers, dates, and other required information. A double bar code system is available, with matching bar codes printed on the label and the leaflet, to reduce the risk of mismatched labelling. Holograms and tamper-evident strips can also be incorporated. Available configurations include booklet formats, concertina folded leaflets, cross-fold leaflets with dual openings, and a number of laminated variations.

The most recent product, called Secure Reseal, is a permanently sealed laminated version with a double-zip perforation. This design ensures that the leaflet doesn't open when applied to tightly curved containers. An easy-lift tab allows users access to the leaflet, which can be repeatedly consulted and resealed.

INK-JET PRINTER

New printhead design optimizes markings: A continuous ink-jet (CIJ) printer features a printhead design with optimized geometry and integral temperature sensors that enable more-precise printing in varying climatic conditions. In addition, the Linx 6800 incorporates an icon-based user interface and improved printer control software, and it is housed within a sturdy IP 65–rated stainless-steel or IP 66–rated plastic cabinet.

Fig. Continuous Ink-jet

The machine, which was introduced by Linx Printing Technologies plc (St. Ives, Cambs, UK) at Emballage, represents the "most fully equipped CIJ printer available in its class," according to sales and marketing director David Jenkins. The modified printhead enables more consistent and accurate ink-drop placement, while a built-in temperature compensation system ensures that print quality does not vary with the ambient temperature. The operator interface includes a black-on-white VGA screen for easy viewing under all lighting conditions, and the new printer control software contains several built-in message formats. Users can select character heights from 1.7 to 13.2 mm and message formats up to 34 drops high.

The Linx 6800 integrates easily with other systems and equipment on the production line, and it prints on a wide variety of substrates.

PACKAGING EQUIPMENT

Multifunction Thermoformer Suited for Sterile Medical Applications

Designed for medium-output requirements, a modular

thermoforming machine is suited for a variety of uses, including the processing of sterile medical equipment. Multivac AG presented the new features of its compact rollstock machine, the R330, at the recent Emballage exhibition.

Side profiles and cross sections as well as the feet and control cabinet are constructed of corrosion-resistant stainless steel. The hermetically sealed cabinet and machine drives are designed to prevent moisture ingress.

In addition, all bearing points are sealed and fitted with maintenance-free slide bearings. Consequently, the thermoformer achieves excellent hygiene standards. Because there are no cavities where water or product can collect, it can be foam cleaned and then simply hosed down. The R330 complies fully with the EC Machinery Directive and meets US FDA requirements for equipment used in the processing of medical devices.

Flexibility is intrinsic to the machine design, according to the company. Two complete die sets can be mounted, and the changeover can be programmed and operated using the appropriate packaging menu. The die's lifting and closing mechanism is controlled by pneumatic cylinders that consume low levels of compressed air; it supports rapid die lifting movements and accelerated adjustments to varying thicknesses of packaging materials.

VACUUM PUMPS

Oil-Free Pumps Reduce Contamination Risks

Oil-free positive-displacement pumps introduced at Medica by Leybold Vakuum GmbH (Köln, Germany) feature clean evacuation without running the risk of hydrocarbon contamination. According to company spokesman Otmar Reuter, the EcoDry L pumps meets the performance specifications of conventional oil-sealed pumps while eliminating potential contamination of the vacuum system or environment.

Because the device does not require periodic changes of pump fluids, traps, or filters, operating costs have been reduced, he adds. Permanently lubricated and sealed bearings located outside of the vacuum unit ensure maintenance-free operation for up to two years. Piston operation was selected because of its mechanical simplicity, reliability, and low cost; the elimination of consumables also contributed to a reduction in operating costs.

Fig. Oil-free Pumps

EcoDry L pumps also feature high pumping speeds at low pressure, minimal noise and vibration, and a low nominal speed of 1100 rpm. The vacuum pumps are suited for such applications as inspection tools, mass spectrometers, sterilizers, electron microscopes, lighting units, thin-film coaters, transfer chambers, and load-lock chambers. They are also appropriate for use as dry prevacuum pumps in high-vacuum systems.

SERVOAMPLIFIER

Euro-Card-Sized Unit Drives Motors up to 500W

Designed to maintain a stable predetermined shaft speed independent of load fluctuations, a four-quadrant servoamplifier drives permanent-magnet dc motors with output power ranging from 80 to 500 W. The ADS E50/10, which supplies 10-A continuous and 20-A maximum output currents, was exhibited by Interelectric AG (Sachseln, Switzerland) at Medica.

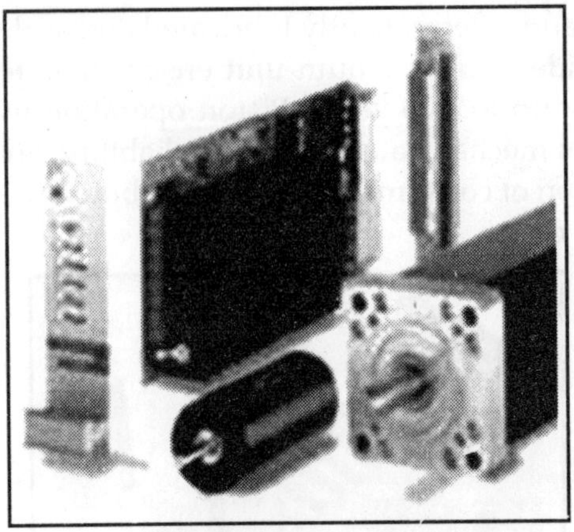

Fig. Euro-card-sized

Four operation modes can be selected by means of a dip switch: I x R compensation, dc tachometer control, digital encoder control, and current control. The amplifier is protected against over-current, overtemperature, and short circuits; a wide range of power inputs enable its use with various power supplies. The unit's Euro-card dimensions enable its installation in a 19-in. subrack or plug-in card system.

At the heart of the company's motor systems, notes marketing director Albert Bucheli, is the proprietary moving-coil rotor. Mass inertia is uncharacteristically low thanks to the low-weight rotor, he says, resulting in high acceleration levels. There is no cogging, and electromagnetic interference is minimal.

MOTOR

Microdrive System Developed for Miniaturized Applications

A microdrive system measuring 1.9-mm-diam was introduced by Dr. Fritz Faulhaber GmbH Company KG and Minimotor S.A. at Medica. The objective in developing this brushless dc motor, according to the company, was to open up new applications that were out of reach because of a lack of suitable subminiature drive components.

Application areas include minimally invasive diagnostic instruments, microsensor assemblies, miniature pumps and grippers, and microactuators for measurement and assembly engineering.

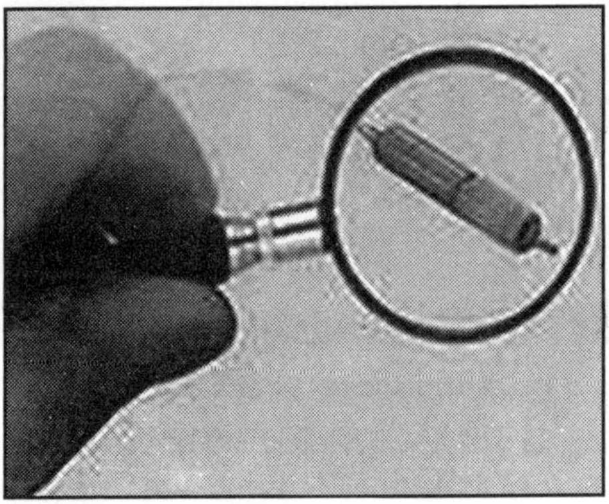

Fig. Microdrive System

The motor operates according to the skew wound-coil principle. The gear head features a reduction ratio of 3.6:1 per gear stage, providing the three-stage gearmotor combination with high torque. Indeed, its torque capability exceeds its own moment of friction. The microdrive is produced under cleanroom conditions.

An external open-loop speed controller operates the microdrive. A three-phase sinusoidal ac signal ensures constant running across the entire speed range up to 100,000 rpm. Operating parameters can be entered via potentiometer and switch or controlled by means of the RS-232 interface.

SCANNER

Ultracompact bar Code Reader Attains 500 Scans Per Second

A 50 × 40 × 28-mm bar code scanner is reportedly the smallest of its type. Introduced at Emballage by Datalogic S.p.A. (Bologna, Italy), the DS2200 incorporates a miniaturized optical laser system developed by the company that enables the overall size reduction. Despite its small size, the component achieves 500 scans per second and 0.15-mm resolution, according to marketing coordinator Jennifer L. Smith.

Fig. Ultracompact Bar Code Reader

Weighing 170 g, the DS2200 scanner also fea-tures a 50–200-mm reading distance. It is suited for machine vision operations in the pharmaceutical sector and for use in clinical analysis applications.

ROTARY PRINTER

Equipment Eliminates Need for Preprinted Rollstock

Fig. Rollstock Packaging Equipment

Designed for use with rollstock packaging equipment, a rotary printer can process medical paper, Tyvek, aluminium, and polyamide/polyethylene foil at speeds up to 25 cycles per minute. Amaco GmbH (Dornbirn, Austria) presented the R 400 at Medica.

"The concept," says sales manager Christoph Lajda, "is to use the blank top web for printing and thereby eliminate the need for various preprinted packaging materials. A variety of logos, bar codes, and so forth can be printed simply by changing the photopolymer cliché." Not only does the R 400 speed the printing process while producing opti-mal quality characters, adds Lajda, but it cuts back significantly on waste packaging material.

STERILE PACKAGING

Multilayer Film Maintains Mechanical Strength After Thermal Exposure

Unlike conventional films that may crystallize following exposure to high temperatures dur-ing sealing and sterilization processes, a multilayer film presented by Wipak Medical (Nastola, Finland) at Medica maintains its mechanical strength and provides a clean peel.

Steriking Multi-X film comprises a heat-resistant outer layer of polyester and intermediate and sealing layers of polypropylene that offer advanced sealing and peeling properties. The seal is sufficiently strong to withstand autoclave sterilization processes, yet offers a clean, fibre-free peel. In addition, the film's green tint grows darker when sealed, providing a visual control that the sealing area is free of defects. The packaged item is easily recognizable through the transparent film.

Microtechnology for Cell Handling and Analysis

Microtechnology is technology with features near one micrometer (one millionth of a meter, or 1ìm). Not only electrical devices, but also mechanical devices, can be miniaturized and batch-fabricated, promising the same benefits to the mechanical world as integrated-circuit (IC) technology has given to the electrical world.

Microfluidics deals with the behaviour, precise control and manipulation of fiuids that are geometrically constrained to a small, typically sub-millimeter, scale.

The advantage is a precise control (temperature and volume) and dosing of the fluid, and low energy and reagent consumption. Microfluidics is, e.g., used in the development of inkjet printheads, genetic test chips, or lab-on-a-chip technology.

Microfluidics for Immobilization and Dosing

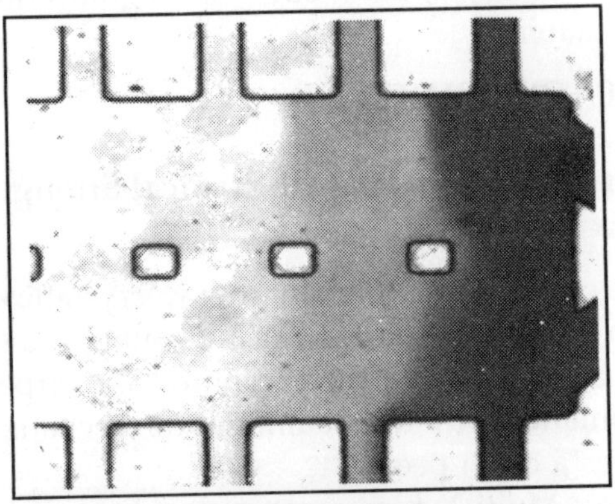

Fig. Microfluidics

Fig. Cell Immobilization

Cells are immobilized on a hole plate via suction, then agents are dosed via a fluidic manifol

Biological Impedance Spectroscopy: Lock-In Amplifier

Impedance spectroscopy is a method for analyzing biological cells and is used in other bio-impedance applications such as electrical impedance tomography (EIT). A key advantage is that no additional preparation or conditioning of the biological cells is necessary, in contrast to many optical analysis methods. Through miniaturization of the measurement setup the necessary amount of analyte can also be greatly reduced, which is another important benefit.

The figure below shows how the impedance changes, when a cell passes the electrodes in the microchannel.

Two electrodes are placed at the ceiling and two at the bottom of the channel. The electrodes at the ceiling are then excited with a sinusoidal voltage, and the current is measured differentially at the bottom electrodes. An impedance change results, which is shown below the channel schematic and is entitled "Cell signal".

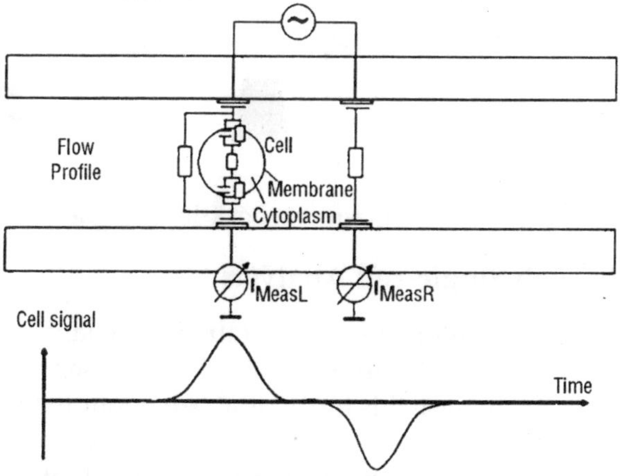

Fig. Flow-Through Cell Impedance Measurements

Impedance changes as a cell passes the electrodes in a microchannel. A lock-in amplifier has been used to perform the impedance measurements. Since it can measure both, amplitude and phase deviation of the input signal, it is an optimal choice for impedance spectroscopy measurements of cells. The figure below explains the functional principle of a lock-in amplifier.

The sensor is supplied with a periodic signal, called local oscillator (L.O), that modulates the sensor signal with a defined frequency. This modulated signal is then amplified, optionally filtered, and then demodulated with the same L.O. This allows for filtering out unwanted disturbances and noise (e.g. drift and 1/f noise) of all frequencies other than the modulation frequency. The lock-in amplifier can be realized in CMOS-technology, which will enable miniaturization and a further integration of the whole measurement system.

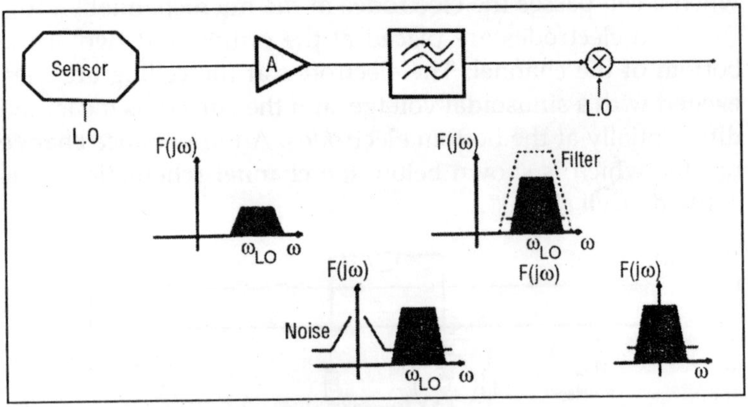

Fig. Block Diagram of a Lock-In Amplifier

Flow-through Single-cell Analysis

The complex interaction of metabolic mechanisms within a single can lead to very individual cell responses to an event or stimulus. To analyse this individual behaviour in the context of Systems Biology it is necessary to combine analytical chemistry methods with new instrumentation that allows for investigating the metabolome of a single cell.

A major challenge is the minute volume of a single cell, which is in the range of one picoliter for a yeast cell. Therefore, a microfluidic to handle single cells coming from, e.g., a cell reactor, has been developed. It includes detectors to monitor cell properties, like the cell size and fluorescence intensity, while entering the microsystem. Subsequently, the cells are lysed, and the metabolites of each single cell are extracted and handed over to a MALDI-TOF mass spectroscopy substrate in a continuous process.

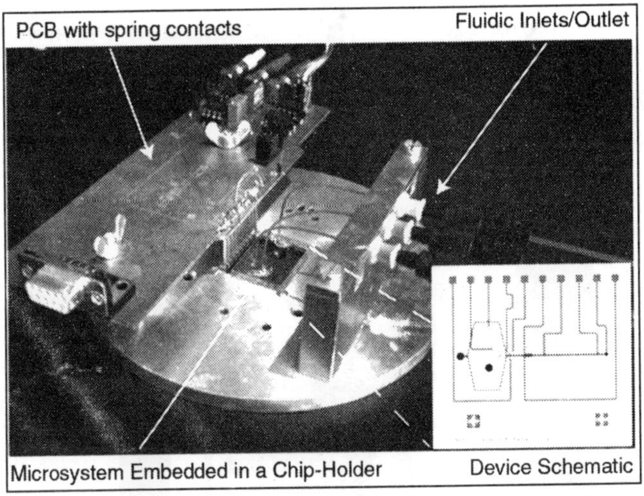

Fig. Single-cell Analysis Microsystem in a Chip-holder

The setup additionally includes a PCB board, spring contacts and fluidic inlets/outlets. It was designed to fit on an inverted microscope.

A glass/polydimethylsiloxane (PDMS) microsystem has been used that consists of three layers (glass, PDMS foil, PDMS block). The chip-holder allows for interfacing the microfluidic device with external electronics and fluidics. It is designed to fit on an inverted light microscope. A stack of two Peltier-elements was used for controlled cooling (stopping of the metabolism upon cell entrance in the microsystem).

HOW THE TECHNOLOGY WORKS

MTI Micro's Mobion technology simplifies the traditional approach to system design in which water would need to be externally pumped from the cathode to the anode side for the chemical reaction to occur and energy to be produced. Using MTI Micro's proprietary technology the water required for the process on the fuel side is transferred internally within the fuel cell from the site of water generation on the air-side of the cell.

This internal flow of water takes place without the need for any pumps, complicated re-circulation loops or other micro-plumbing tools. Below are two schematics of typical direct methanol fuel cell system designs and a schematic of MTI Micro's Mobion technology

Low Energy Content Direct Methanol Fuel Cell (DMFC) Active System

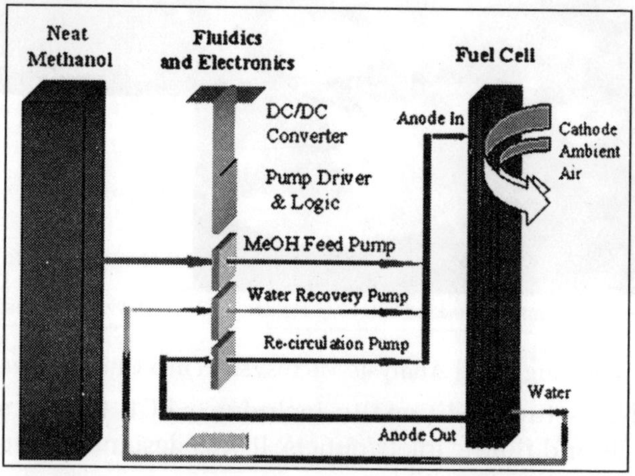

Fig. Anode Fuel Concentration

Typically, anode fuel concentration is controlled at a low methanol level (2 per cent) and water is needed at the anode side for the reaction to occur so the water is collected from the cathode side (e.g. by gravity) and pumped externally back to the anode. The result is a system that uses micro-pumps and micro-pipes increasing the size, complexity and cost of the system and ultimately impacting commercial product requirements.

Low Energy Content Direct Methanol Fuel Cell (DMFC) Passive System

Another direct methanol fuel cell (DMFC) system design

- for a low content DMFC- carries the water required for the reaction to occur at the anode side together with the fuel in the same tank, thereby avoiding the need of collecting and pumping water from the cathode back to the anode.

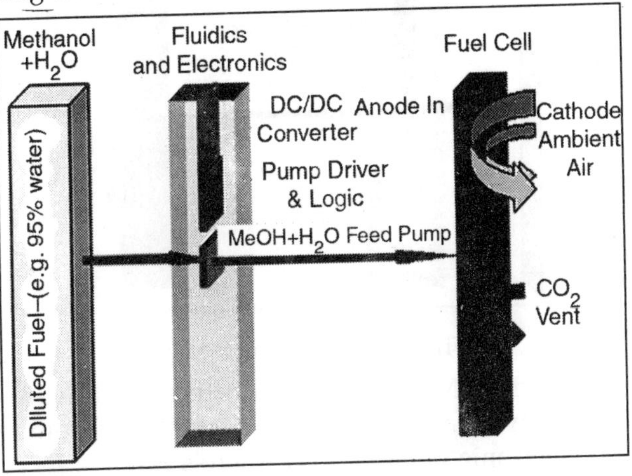

Fig. Direct Methanol Fuel Cell

In this system design, carrying water with the fuel severely reduces the system's energy density because water has no energy content. The result is a simplified system that might not improve the energy density needs of today's portable electronic devices and may not compete favorably against advanced lithium ion batteries.

MTI Micro Mobion Technology

In repeated laboratory tests, MTI Micro's Mobion technology allows for neat (100 per cent) methanol to be directly fed into the anode side of the direct methanol fuel cell (DMFC), consequently eliminating the need for carrying water in the system or the need for micro-pumps and a micro-plumbing subsystem.

While not shown in detail for reasons of remaining need of confidentiality, the important features of such an approach are how to properly control supply of 100 per cent methanol to the cell with uniform distribution across the cell, and how to achieve water flow within the cell from the cathode (air) side to the anode (fuel) side without the use of a pump.

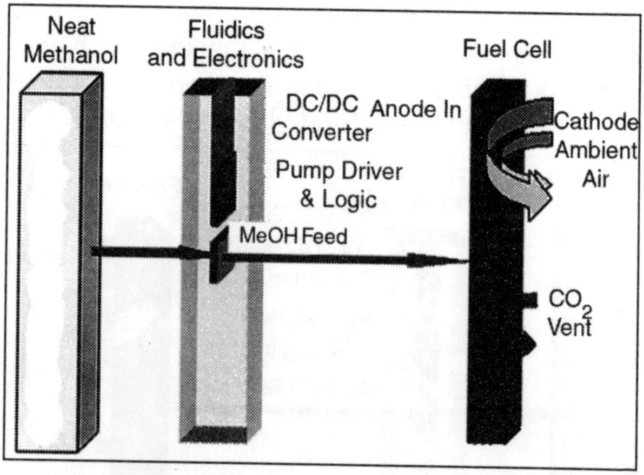

Fig. Micro's Mobion System

MTI Micro's Mobion system offers the advantage of higher energy density than the current technologies for portable electronic devices. MTI Micro's Mobion technology is designed to reduce parts count and the need for complex components resulting in a smaller system. By developing this very simple, complete system, manufacturability needs and market opportunities for small fuel cells could be addressed early and cost effectively.

The system is also scalable to a wide range of product options - from accessories to battery replacements - in both the commercial and military markets. A series of system prototypes have been developed showing size reductions and performance improvements, including operation on an increased concentration of up to 100 per cent methanol.

Chapter 7

Microbiology

Microbiology could be defined as the study of organisms too small to be seen with the naked eye. However, the recent discovery of bacteria of near 1 mm in size has made this definition somewhat inaccurate and in the grand tradition of science, a new definition is in order.

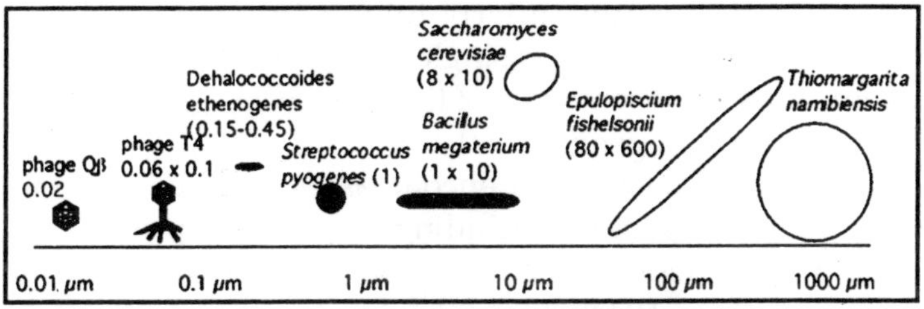

Fig. The Relative Size of Microbes

We will consider microbiology to be the study of organisms that can exist as single cells, contain a nucleic acid genome for at least some part of their life cycle, and are capable of replicating that genome.

This broad description encompasses an understandably large group of organisms including fungi, algae, protozoa and bacteria. This definition would also include viruses, which microbiology texts traditionally discuss along with living organisms.

Microbiology also involves a collection of techniques to study and manipulate these small creatures. Because of their size, special instruments and methods had to be developed to allow the performance of interpretable experiments on microorganisms. These methods are not restricted to microbes alone, but have also found utility in working with populations of cells from higher organisms.

With apologies to other small organisms, this book will mostly focus on bacteria (which we will also call microorganisms or microbes) and their impact on the rest of the biosphere. This can be weakly justified by the fact that bacteria have a major impact on the world around us and, because of their perceived importance, more research and knowledge has been accumulated about them.

Microorganisms are everywhere, but why are they worth learning about? The short answer is that they affect our life in many different ways. Before we begin our study of these creatures, we will first take a tour of some of their important habitats and point out why our existence depends upon them. We will then briefly explore the history of microbiology.

We will consider microbiology to be the study of organisms that can exist as single cells, contain a nucleic acid genome for at least some part of their life cycle, and are capable of replicating that genome.

This broad description encompasses an understandably large group of organisms including fungi, algae, protozoa and bacteria. This definition would also include viruses, which microbiology texts traditionally discuss along with living organisms.

Microbiology also involves a collection of techniques to study and manipulate these small creatures. Because of their size, special instruments and methods had to be developed to allow the performance of interpretable experiments on microorganisms. These methods are not restricted to microbes alone, but have also found utility in working with populations of cells from higher organisms.

With apologies to other small organisms, this book will mostly focus on bacteria (which we will also call microorganisms or microbes) and their impact on the rest of the biosphere. This can be weakly justified by the fact that bacteria have a major impact on the world around us and, because of their perceived importance, more research and knowledge have been accumulated about them.

Microorganisms are everywhere, but why are they worth learning about? The short answer is that they affect our life in many different ways. Before we begin our study of these creatures, we will first take a tour of some of their important habitats and point out why our existence depends upon them. We will then briefly explore the history of microbiology.

MICROBES HAVE A LARGE IMPACT ON HUMAN HEALTH

While death from infectious disease in many countries has been greatly diminished, infection rates in developing nations remain unacceptably high. "Ancient" diseases continue to be a problem where nutrition and sanitation are poor, and emerging diseases such as Acquired Immunodeficiency Syndrome (AIDS) are even more dangerous for such populations. The Centers for Disease Control and Prevention (the U.S. government agency charged with protecting the health and safety of people) estimates that about 9 per cent of adults between the ages of 18-49 in Sub-Saharan Africa are infected with HIV.

AIDS is only one of a number of new diseases that have emerged. Many of these diseases have no known cure. Influenza and pneumonia are leading killers of the elderly even in the U. S. and other developed nations. Even the common cold causes illness and misery for almost everyone and drains the productivity of all nations. Many of the new diseases are viral in nature, making them notoriously difficult to treat.

Disease due to food-borne pathogens also remains a problem, largely because of consumption of improperly processed or stored foods. Understanding the sources of contamination and developing ways to limit the growth of pathogens in food is the job of food microbiologists.

Surprisingly, many diseases that were previously thought to have only behavioral or genetic components have been found to involve microorganisms. The clearest case is that of ulcers, which was long thought to be caused by stress and poor diet. However the causative agent is actually a bacterium, *Helicobacter pylori*, and many ulcers can be cured with appropriate antibiotics. Work on other non-infectious diseases such as heart disease, stroke and some autoimmune diseases also suggest a microbial component that triggers the illness.

Finally, some pathogenic microbes that had been "controlled" through the use of antibiotics are beginning to develop drug resistance and therefore reemerge as serious threats in the industrialized world as well as developing nations. Tuberculosis is an illness that was on the decline until the middle 80's.

It has recently become more of a problem, partly due to drug resistance and partly due to a higher population of immunosuppressed individuals from the AIDS epidemic. *Staphylococcus aureus* strains are emerging that are resistant to many of the antibiotics that were previously effective against them. These staph infections are of great concern in hospital settings around the world. Understanding both familiar killers and new pathogens will require an understanding of their biology, and thus an understanding of the field of microbiology.

MICROBES ARE OFTEN HELPFUL, NOT HARMFUL

From the beginning of microbiology, significant resources have been spent to understand and fight disease-causing microorganisms. We may be surprised to learn that only a small fraction of microbes are involved in disease; many other microbes actually enhance our well-being. The harmless microbes that live in our intestines and on our skin actually help us fight off disease.

They actively antagonize other bacteria and take up space, preventing potential pathogens from gaining a foothold on our bodies. Indeed, like all large organisms, humans have entire communities of microorganisms in their digestive systems that contribute to their overall health. The microbial community in humans not only protects us from disease, but also provides needed vitamins, such as B_{12}.

Human health and nutrition also depends on healthy farm animals. Cows, sheep, horses and other ruminant animals utilize their microbial associates to degrade plant material into useful nutrients. Without these bacteria inside ruminants, growth on plant material would be impossible.

Commercial crops are also central to human prosperity, and much of agriculture depends upon the activities of microbes. For example, an entire group of plants, the legumes, forms a cooperative relationship with certain bacteria. These bacteria convert nitrogen gas to ammonia for the plant, an important nutrient that is often limiting in the environment. Microbes also serve as small factories, producing valuable products such as cheese, yogurt, beer, wine, organic acids and many other items. In conclusion, while it is less apparent to us, the positive role of microbes in human health is at least as important as the negative impact of pathogens.

MICROBIAL EFFECT ON THE ENVIRONMENT

Microbes make up the major portion of the biomass present on the Earth. Therefore, the nutrients they eat and the products they form greatly influence the environment. Cyanobacteria and algae in the oceans are responsible for most photosynthesis and are a major sink for carbon dioxide, a greenhouse gas.

Microbes release nutrients from dead organisms, making them available to the rest of the ecosystem. Some microbes play a role in the production of energy, while other microbes interfere with energy production.

Whether measured by the number of organisms or by total mass, the vast majority of life on this planet is microscopic. These teaming multitudes profoundly influence the make-up and character of the environment in which we live. Presently, we know very little about the microbes that live in the world around us because less than 2 per cent of them can be grown in the laboratory. Understanding which microbes are in each ecological niche and what they are doing there is critical for our understanding of the world.

Microbes are the major actors in the synthesis and degradation of all sorts of important molecules in environments. Cyanobacteria and algae in the oceans are responsible for the majority of photosynthesis on Earth. They are the ultimate source of food for most ocean creatures (including whales) and replenish the world's oxygen supply. Cyanobacteria also use carbon dioxide to synthesize all of their biological molecules and thus remove it from the atmosphere. Since carbon dioxide is a major greenhouse gas, its removal by cyanobacteria affects the global carbon dioxide balance and may be an important mitigating factor in global warming.

In all habitats, microorganisms make nutrients available for the future growth of other living things by degrading dead organisms. Microbes are also essential in treating the large volume of sewage and wastewater produced by metropolitan areas, recycling it into clean water that can be safely discharged into the environment.

Less helpfully (from the view of most humans), termites contain microorganisms in their guts that assist in the digestion of wood, allowing the termites to extract nutrients from what would otherwise be indigestible. Understanding of these systems helps us to manage them responsibly and as we learn more we will become ever more effective stewards.

Energy is essential for our industrial society and microbes are important players in its production. A significant portion of natural gas comes from the past action of methanogen (methane-producing bacteria). Numerous bacteria are also capable of rapidly degrading oil in the presence of air and special precautions have to be taken during the drilling, transport and storage of oil to minimize their impact.

In the future, microbes may find utility in the direct production of energy. For example, many landfills and sewage treatment plants capture the methane produced by methanogens to power turbines that produce electricity. Excess grain, crop waste and animal waste can be used as nutrients for microbes that ferment this biomass into methanol or ethanol. These biofuel are presently added to gasoline and thus decreasing pollution. They may one-day power fuel cells in our cars, causing little pollution and having water as their only emission.

Finally, We are increasingly taking advantage of the versatile appetite of bacteria to clean up environments that we have contaminated with crude oil, polychlorinated biphenyls (PCBs) and many other industrial wastes. This process is termed bioremediation and is a cheap and increasingly effective way of cleaning up pollution.

STUDYING MICROBES HELPS US TO UNDERSTAND THE WORLD AROUND US

Microorganisms used in research have many useful properties. They will grow on simple, cheap medium and will often rise to large populations in a matter of 24 hours. It is easy to isolate their genomic material, manipulate it in the test tube and then place it back into the microbe. Due to their large populations it is possible to identify rare events and then, with the use of powerful selective techniques, isolate interesting bacterial cells and study them. These advantages have made it possible to test hypothesis rapidly. Using microbes scientists have expanded our knowledge about life.

Microorganisms have been indispensable instruments for unlocking the secrets of life. The molecular basis of heredity and how this is expressed as proteins was described through work on microorganisms. Due to the similarity of life at the molecular level, this understanding has helped us to learn about all organisms, including ourselves.

Some prokaryotes are capable of growing under unimaginably harsh conditions and define the extreme limits of where life can exist. Some species have been found growing at near 100 °C in hot springs and well above that temperature near deep-sea ocean vents. Others make their living at near 0 °C in freshwater lakes that are buried under the ice of Antarctica. The ability of microbes to live under such extreme conditions is forcing scientists to rethink the requirements necessary to support life. Many now believe it is entirely possible that Jupiter's moon Europa may harbor living communities in waters deep below its icy crust. What may the rest of the universe hold?

Until recently, while we could study specific types of bacteria, we lacked a cohesive classification system, so that we could not readily predict the properties of one species based on the known properties of others. Visual appearance, which is the basis for classification of large organisms, simply does not work with many microbes because there are few distinguishing characteristics for comparison between species even under the microscope.

However, analysis of their genetic material in the past 20 years has allowed such classification and spawned a revolution in our thinking about the evolution of bacteria and all other species. The emergence of a new system organizing life on Earth into three domain is attributable to this pioneering work with microorganisms.

The fruits of basic research on microbes has been used by scientists to understand microbial activity and therefore to shape our modern world. Human proteins, especially hormone like insulin and human growth factor, are now produced in bacteria using genetic engineering. Our understanding of the immune system was developed using microbes as tools. Microorganisms also play a role in treating disease and keeping people healthy. Many of the drugs available to treat infectious disease originate from bacteria and fungi.

One last recent role of microbes in informing us about our world has been the tools they provide for molecular biology. Enzymes purified from bacterial strains are useful as tools to perform many types of analyses. Such analyses allow us to determine the complete genome sequence of almost any organism and manipulate that DNA in useful ways.

We now know the entire sequence of the human genome, with the exception of regions of repetitive DNA, and this will hopefully lead to medical practices and treatments that improve health. We also know the entire genome sequence of many important pathogens. Analysis of this data will eventually lead to an understanding of the function of critical enzymes in these microbes and the development of tailor-made drugs to stop them.

The tools of molecular biology will also affect agriculture. For example, we now know the complete genome sequence of the plant *Arabidopsis* (a close relative of broccoli and cauliflower). This opens a new avenue to a better understanding of all plants and hopefully improvements in important crops.

Microbes have a profound impact on every facet of human life and everything around us. Pathogens harm us, yet other microbes protect us. Some microbes are pivotal in the growth of food crops, but others can kill the plants or spoil the produce. Bacteria and fungi eliminate the wastes produced in the environment, but also degrade things we would rather preserve. Clearly they effect many things we find important as humans.

MICROBIOLOGY IS A WEB OF DISCOVERIES

Before we begin the adventure that we call learning microbiology, a look at the history of microbiology will help us to understand the contributions of those who have come before. This perspective will hopefully give us an appreciation of their efforts and put the body of knowledge we will examine in the context of history.

Keep in mind that microbiology is a relatively young science. It was only 130 years ago that it became possible to seriously study microorganisms in the laboratory, with most of our understanding of microbes coming in the last 60 years.

The history of microbiology, like all human history, is not a catalog of linear progress, but is more of an interweaving of the careers of bright individuals and their insights. Each new discovery relied on previous ones and in turn spawned further inquiry. A web of interdependent concepts evolved over time through the work of scientists in many related disciplines and nations. Often the research of one individual impacted the efforts of another studying a completely different problem. Keep this in mind as we look at this history.

Below we present several journeys through this web, mentioning some individuals who were particularly important in the progression. This history reflects our view of important events of the past, but is by no means comprehensive.

We will first look at the development of the techniques for handling microorganisms, since everything else in microbiology depends upon these procedures. Next, we will examine how these techniques helped to settle an old debate, the question of spontaneous generation. Then, we will look at the history of infectious disease.

The science of microbiology had its most significant early impact on human health, uncovering the cause of the major killers of the day, and then methods to treat them. As microbiology matured, scientists began to look at what non-pathogenic microbes were doing in the environment and we will look a bit at the history of general microbiology.

MICROSCOPES AND DISCOVERY OF MICROBES

For years the existence of microorganisms was suspected, but could not be proven, since bacteria were too small to be seen with the naked eye. It took the microscope to expose their tiny world and that instrument has been linked to microbiology ever since. In 1664, Robert Hooke devised a compound microscope and used it to observe fleas, sponges, bird feathers, plants and molds, among other items. His work was published in *Micrographia* and became a popular and widely read book at the time.

Several years later Anton van Leeuwenhoek, a fabric merchant and amateur scientist (or "natural philosopher" as such people called themselves), became very adept at grinding glass lenses to make telescopes and microscopes. While crude by modern standards, his were a technical marvel for the time, able to magnify samples greater than 200-fold - a reproduction of a Leeuwenhoek microscope.

They also produced clearer images than the compound microscopes of the time. By peering through his microscope, Leeuwenhoek observed tiny organisms or "wee animacules" as he called them.

He spent months looking at every kind of sample he could find and eventually submitted his observations in a letter to the Royal Society of London, causing a sensation. Hooke was asked to confirm the findings of Leeuwenhoek and his affirmative assessment garnered them wide acceptance. Surprisingly the work of these two scientists was not followed up for almost 200 years. Human societies had neither the technical prowess nor the inclination to develop the science of microorganisms. It was not until the rise of the industrial revolution that governments and people dedicated the financial and physical resources to understand these small inhabitants of our world.

With the development of better microscopes in the 19th century, scientists returned to an examination of microorganisms. After finishing his education, Ferdinand Julius Cohn was able to convince his father to lay down the large sum necessary to purchase a microscope for him, one better than that available at the University in Breslau, then part of Germany. He used it to carefully examine the world of the microbe and made many observations of eukaryotic microorganisms and bacteria.

His landmark papers on the cycling of elements in nature was published in *Ueber Bakterien* in 1872 and a microbial classification scheme including descriptions of *Bacillus* were published in the first volume of a journal he founded, *Beitraege zur Biologie der Planzen*. Cohn's work with microscopes popularized their use in microbiology. This and his other work inspired many other scientists to examine microbes. Cohn's encouragement of Robert Koch, a German physician by training, began the field of medical microbiology.

MICROBIOLOGY TECHNIQUES

Robert Koch, searched for the causes of many diseases. Through these investigations he and his laboratory developed many classic microbial techniques. He used adaptations of the staining methods of Carl Weigert to begin the process of distinguishing microbes and identifying pathogens. His lab was the first to isolate a disease-causing organism.

A major contribution to bacterial techniques was the development of methods using solid medium for the cultivation of bacteria. Koch was convinced that microbes caused some diseases.

However, to test this idea, he needed to isolate the causative agent. Almost all samples from diseased animals or any natural surface contained many different microbes and it was impossible to tell which one was the problem. A method was needed to separate these different bacteria.

The most common method of isolation was to continually dilute a sample in liquid broth in hopes that at high enough dilution, only one type of microbe would be found. A problem with this method is that only the most populous microbe would be isolated, but that might not be one causing the disease. There were other technical problems as well with such a liquid-based system, so a solid medium would seem to provide distinct advantages.

Koch had tried gelatin for these experiments with unsatisfactory results. Building on the work of Brefeld and Schroeter, Koch used potato slices as a solid medium and observed that a boiled potato left in the open air would develop tiny circular raised spots.

Examination of these spots revealed they were made up of microorganisms and each spot had just one type of microbe in it. He realized that these colonies were pure cultures of bacteria and probably arose from a single species of microbe from the air that landed on the potato. By boiling a potato, slicing it with a hot knife and keeping it in a sterile container with a lid, Koch could keep the potato sterile. But if a sample from a disease animal was smeared across the potato, colonies arose, each being pure isolates from the animal. By then testing these isolates in animals, Koch was able to isolate the cause of anthrax, *Bacillus anthracis*.

Potatoes failed to support the growth of many microorganisms and Koch and his laboratory were constantly frustrated by the lack of a good solid medium. Walter Hesse joined Koch's laboratory to do studies on air quality, showing a remarkable attention to detail and patience in his work.

His wife, Angelina Fannie Hesse along with raising their three sons, also would assist her husband with his research in the laboratory. Walter was attempting to do his air quality experiments using medium containing gelatin as the solidifying agent. In the summertime, temperatures would often rise above the melting point of gelatin, then used as the solidifying agent. In addition, microbes would often grow in the cultures that were capable of degrading gelatin and in both cases this would cause liquefaction of the medium, ruining the experiments.

One day while eating lunch, the frustrated scientist asked Lina (as she was called) why her jellies and puddings stayed solid even in the hot summer temperatures. She told him about agar-agar, a heat resistant gelling agent that she had learned about while growing up in New York from a Dutch neighbour who had emigrated from Java.

Development of the new agent by Angelina and Walter led to a resounding success. Few microbes are able to degrade agar and it melts at 100 °C yet remains molten at temperatures above 45 °C. This allows the mixing of the agar with heat-sensitive nutrients and microbes.

After solidification, it will not melt until a temperature of 100 °C is again attained, facilitating the easy cultivation of pathogens. It can also be stored for long periods of time, allowing the cultivation of slow-growing microorganisms. Any type of broth can be mixed with agar, giving great flexibility in the kinds of medium that can be made. Thus, many more types of microbes could be cultivated.

Koch's laboratory also developed methods of pure culture maintenance and aseptic technique. Aseptic technique involves the manipulation of pure cultures in a manner that prevents their contamination by outside microorganisms. Equally important, aseptic technique prevents their spread into the environment.

Remember that Koch was studying some of the most devastating microbial pathogens of the period, and their release could potentially cause disease in the scientists working on them. These procedures were also absolutely critical because they allowed careful study of pure microorganisms, making it possible to identify the role of each microbe in a given situation.

Another problem in the cultivation of microbes was solved by Julius Petri while working in Koch's laboratory. Solid medium was poured on glass plates and allowed to spread and harden. Once cooled it allowed a solid surface for streaking. However, creation of these plates required great care since exposure to the air (and the microbes in it) often leads to contamination. In addition, to prevent contamination of plates during incubation, a cumbersome bell jar was used.

If one wanted to view samples, the plate had to be removed from the jar, further exposing it to unwanted microbes in the air. In 1887 Petri developed shallow glass dishes, with one having a slightly larger diameter than the other. Medium is poured into the smaller dish and the larger one serves as a cover. This simple device solved all of the above problems and took on the name of its inventor, the petri plate.

These same techniques are essential in studying all microorganisms. Collectively the above techniques have been used to isolate and identify thousands of different microorganisms. As a testament to the significance of their achievement, these techniques are practiced with remarkably little change in every laboratory that works with microorganisms today.

SPONTANEOUS GENERATION

- For many centuries many people believed in the concept of spontaneous generation, the creation of life from organic matter.
- Francesco Redi disproved spontaneous generation for large organisms by showing that maggots arose from meat only when flies laid eggs in the meat.
- Spontaneous generation for small organisms again gained favour when John Needham showed that if a broth was boiled (presumed to kill all life) and then allowed to sit in the open air, it became cloudy.
- Louis Pasteur ended the debate with his famous swan-neck flask experiment, which allowed air to contact the broth. Microbes present in the dust were not able to navigate the tortuous bends in the neck of the flask.

Spontaneous generation is the hypothesis that some vital force contained in or given to organic matter can create living organisms from inanimate objects. Spontaneous generation was a widely held belief throughout the middle ages and into the latter half of the 19th century.

In fact, some people still believe in it today. The idea was attractive because it meshed nicely with the prevailing religious views of how God created the universe. There was a strong bias to legitimize the idea because this vital force was considered a strong proof of God's presence in the world. Many recipes and experiments were offered in proof. To create mice, a recipe called for dirty underwear and wheat grain to be mixed in a bucket and left open outside.

In 21 days or less, we would have mice. The real cause may seem obvious from a modern perspective, but to the proponents of this idea, the mice spontaneously arose from the wheat kernels.

Another often-used example was the generation of maggots from meat that was left in the open. The failing here was revealed by Francesco Redi in 1668 with a classic experiment. Redi suspected that flies landing on the meat laid eggs that eventually grew into maggots. Here he used three pieces of meat. One piece of meat was placed under a piece of paper. The flies could not lay eggs onto the meat and no maggots developed.

The second piece was left in the open air, resulting in maggots. In the final test, a third piece of meat was overlaid with cheesecloth. The flies were able to lay the eggs into the cheesecloth and when this was removed no maggots developed. However, if the cheesecloth containing the eggs was placed on a fresh piece of meat, maggots developed, showing it was the eggs that "caused" flies and not spontaneous generation. This helped to end the debate about spontaneous generation for large organisms. However, spontaneous generation was so seductive a concept that even Redi believed it was possible in other circumstances.

The concept and the debate were revived in 1745 by the experiments of John Needham. It was known at the time that heat was lethal to living organisms.

Needham theorized that if he took chicken broth and heated it, all living things in it would die. After heating some broth, he let a flask cool and sit at a constant temperature. The development of a thick turbid solution of microorganisms in the flask was strong proof to Needham of the existence of spontaneous generation.

Lazzaro Spallanzani later repeated the experiments of Needham, but removed air from the flask, suspecting that the air was providing a source of contamination. No growth occurred in Spallanzani's flasks and he took this as evidence that Needham was wrong. Proponents of spontaneous generation discounted the experiment by asserting that air was required for the vital force to work.

It was not until almost 100 years later that the great French chemist Louis Pasteur, put the debate to rest. He first showed that the air is full of microorganisms by passing air through gun cotton filters. The filter trapped tiny particles floating in the air. By dissolving the cotton with a mixture of ether and alcohol, the particles were released and then settled to the bottom of the liquid.

Inspection of this material revealed numerous microbes that resembled the types of bacteria often found in putrefying media. Pasteur realized that if these bacteria were present in the air then they would likely land on and contaminate any material exposed to it.

Pasteur then entered a contest sponsored by The French Academy of Sciences to disprove the theory of spontaneous generation. Similar to Spallanzani's experiments, Pasteur experiment, used heat to kill the microbes, but left the end of the flask open to the air. In a simple, but brilliant modification, the neck of the flask was heated to melting and drawn out into a long S-shaped curve, preventing the dust particles and their load of microbes from ever reaching the flask. After prolonged incubation the flasks remained free of life and ended the debate for most scientists.

Pasteur filled a flask with medium, heated it to kill all life, and then drew out the neck of the flask into a long S. This prevented microorganisms in the air from entering the flask, yet allowed air to flow freely. If the swan neck was broken, microbes could enter the flask and grow.

A final footnote on the topic was added when John Tyndall showed the existence of heat-resistant spore in many materials. Boiling does not kill these spores and their presence in chicken broth, as well as many other materials, explains the results of Needham's experiments.

While this debate may seem silly from a modern perspective, remember that the scientists of the time had little knowledge of microorganisms. Koch would not isolate microbes until 1881.

The proponents of spontaneous generation were neither sloppy experimenters nor stupid. They did careful experiments and interpreted them with their own biases. Detractors of the theory of spontaneous generation were just as guilty of bias, but in the opposite direction. In fact, it is somewhat surprising that Pasteur and Spallanzoni did not get growth in their cultures, since the sterilization conditions they used would often not kill endospores. Luck certainly played a role.

It is important keep in mind that the discipline of science is performed by humans with all the fallibility and bias inherent in the species.

Only the self-correcting nature of the practice reduces the impact of these biases on generally held theories. Spontaneous generation was a severe test of scientific experimentation, because it was such a seductive and widely held belief. Yet, even spontaneous generation was overthrown when the weight of careful experimentation argued against it.

MICROBES ARE DISCOVERED TO CAUSE DISEASE

It was long suspected that living things were the agents of disease. In volume 6 of his epic poem *De Rerum Natural (On the Nature of the Universe)* written sometime around 50 B.C., Titus Lucretius Carus speculates about invisible atoms causing disease. This was only one idea among many and some thought that an imbalance in humors caused illness, while others felt that supernatural forces were at work.

The prevailing theory held by most doctors of the 19th century was that chemical toxins were carried from an ill patient to others, causing them to contract the same malady.

The bacteria that were known to be present were seen as a symptom of the disease and not its cause.

Ignaz Semmelweis, a Hungarian physician working in Vienna, made the first breakthrough in the true nature of disease. He realized that asepsis in obstetrical wards could prevent the transmission of childbirth fever from patient to patient. He therefore instigated a policy for all attending physicians to wash their hands with with chloride of lime (a mixture of calcium chloride hypochlorite, CaCl(OCl); calcium hypochlorite, $Ca(OCl)_2$; and calcium chloride, $CaCl_2$) between patients. This innovation dropped the mortality rate from 18 per cent to 2.4 per cent.

Ignaz became a vigorous proponent of his ideas, but the Hungarian doctor's efforts were opposed by many who could not accept that physicians themselves could be responsible for spreading bacterial infection. Ridicule of his idea caused him to move from Vienna to Pest, Hungary and ultimately played a role in a nervous breakdown.

Ironically Semmelweis died from an infection that he contracted during a surgery he performed, while recovering from his nervous breakdown.

Before his death he published his ideas in a paper The *Cause, Concept, and Prophylaxis of Childbed Fever* in 1861. Although the poor writing of the paper contributed to the obscurity of his ideas, the work was ignored for 17 years, which raises an interesting point about the culture of science. Radical ideas, even those that are correct and can save lives, are sometimes ignored. It takes time to overcome the dogma of the day. The personalities involved and the negative light it might throw on past practices play a large role in the rate of acceptance of a new idea.

In a seemingly unrelated event, Louis Pasteur found something interesting while working on wine souring, a problem where wine fermentations produce a sour taste and very little alcohol. Fermentation of alcoholic beverages was thought to be a simple chemical reaction.

Heating, common in most beverage preparations, was thought to cause the breakdown of sugar into alcohol. Pasteur was asked to help out the wine industry in France because wine souring was pushing it close to ruin. His work on wine-making revealed that the process of converting sugar to alcohol is actually performed by various yeast strains.

He then showed that the wine was going bad because a contaminating microbe was generating lactic acid instead of alcohol from the sugar. This idea was controversial, but gained credibility when Pasteur solved the problem by heating the wine and killing the contaminant.

The heating process was named pasteurization in his honour and is still widely used today. In a brilliant step of generalization, Pasteur realized that souring of wine and infectious disease shared a common thread in that they both might involve infection by a microorganism. His suggestion that microbes cause disease became known as the germ theory of disease.

Joseph Lister became aware of Semmelweis's work and together with Pasteur realized the true nature of disease. He then recognized that he could use this idea to help his surgery patients. At this time, major injuries, broken bones or surgery would often result in infection of the damaged area, sometimes leading to amputation or death.

Lister found he could greatly reduce the number of microorganisms on wounds and incisions by using bandages treated with phenic acid, a compound that killed microorganisms (phenic acid, now known as phenol, is the active ingredient in Listerine).

During surgery he began the practice of spraying the wound with a fine mist of phenic acid to kill microbes. These practices greatly reduced the rate of infection and mortality of surgery patients, lending further credence to the germ theory of disease.

In 1876 Robert Koch provided definitive proof of the germ theory by isolating the cause of anthrax and showing it to be a bacterium. From this came the development of Koch's Postulates, a set of rules for the assignment of a microbe as the cause of a disease:

- The specific organism should be shown to be present

in all cases of animals suffering from a specific disease, but should not be found in healthy animals.

- The specific microorganism should be isolated from the diseased animal and grown in pure culture on artificial laboratory media.
- This freshly isolated microorganism, when inoculated into a healthy nonimmune laboratory animal, should cause the same disease seen in the original animal.
- The microorganism should be reisolated in pure culture from the experimental infection.

The postulates are Koch's most famous contribution to science and it is a testament to the utility of these postulates that they are stilled used today to discover the cause of new emerging diseases. Koch went on to apply these principles in the study of many other diseases including tuberculosis, cholera and sleeping sickness.

It should be pointed out that Koch's postulates cannot be applied to all diseases. For example if a disease-causing microbe has humans as its sole host and has a significant possibility of causing death, it would be unethical to apply this microbe to test humans as dictated by postulate 3. Also, it is not always possible to obtain a disease-causing microbe in pure culture.

While attacking the problem of disease, Koch developed the tools for obtaining pure cultures. Advances in science often come from innovations in the available technology. Robert Koch was an important microbiologist because his pioneering work in the isolation and characterization of bacterial diseases helped to identify the causes of many of the maladies plaguing humanity. Further work by other scientists then began the long road to conquering them.

VIRUSES ARE SHOWN TO PARASITIZE ORGANISMS

The ensuing years brought numerous discoveries about the nature of disease. In the latter half of the 19th century, the causative agents for anthrax, tuberculosis, gonorrhea, diphtheria and many more were discovered. Great strides in understanding maladies caused by bacteria were made at this time, including the solidification of the germ theory of disease. However, some illnesses seemed not to have bacterial origins. Microscopic examination of sera from ill patients revealed no organisms and the causative agents could not be grown on any known medium. Yet if these sera were injected into a susceptible host, disease resulted.

In 1886 John Brown Buist devised a method for staining and fixing liquid from a cowpox vesicle. Observations of this slide showed tiny bodies that he believed were spores. Although he did not realise it, he was the first person to see (and photograph) a virus. It was not until the middle of the twentieth century with the invention of the electron microscope that the true shape and structure of viruses was understood.

In 1884 Charles Chamberland in Pasteur's laboratory created an unglazed porcelain filter that had pores much smaller than bacteria (0.1-1 μm). It was possible to pass a solution containing bacteria through these filters and completely remove them from the solution. This enabled the creation of sterile medium without the use of heat, and also became a standard test for the removal of all bacteria, especially when testing transmission of disease. By removing the microbes, Chamberland clearly demonstrated that an infectious agent, and not the solution, was causing the illness. This concept seemed reasonable until scientists began to investigate a tobacco infection.

Tobacco mosaic disease is characterized by light and dark green areas on plant leaves in a mosaic pattern. The disease stunts the growth of the plant, therefore reducing yields. Europeans recognized the effects of this disease soon after tobacco was introduced from the New World in the 17th century. Adolph Mayer first described transmission of the disease by injecting fluid from a diseased plant into a healthy one. Nine out of ten times the healthy plant would become heavily diseased.

In 1892 Dmitrii Ivanowski extended this research by the shocking discovery that the causative agent could pass through

Chamberland's porcelain filter. He reported his findings, but the idea that the causative agent could pass through the filter was so troublesome to Ivanowsky he attributed the phenomenon to a cracked filter or to small spores that passed through the pores. It was Martinus Beijerinck in 1898 who realized the true nature of these particles, making the intellectual leap that the causative agent of the disease must be so small as to pass through the filter known to trap all bacteria. He coined the term *contagium vivum fluidum* a contagious living fluid.

Some scientists thought these agents were toxins in the fluids of the hosts, but further study revealed that tobacco mosaic disease could cause illness through many transfers from plant to plant. A toxin might cause damage on the first plant, but subsequent transfers should make it so dilute that it would no longer have any effect.

These filterable entities were different from bacteria and appeared to depend on their host in order to multiply. The term ultrafiltrable viruses and later just viruses (virus means poison in Latin) was coined to describe these tiny pathogens.

Bacteria are also vulnerable to viruses. Bacterial viruses were described independently by Frederick Twort in 1915 and Felix d'Herelle in 1917. Felix d'Herelle was studying a plague of locusts in Mexico when he noticed that the locusts were being killed by a microorganism.

During experiments to characterize this microbe, d'Herelle noticed the appearance of clear, circular spots two or three millimeters in diameter on cultures growing on agar. At this point d'Herelle studied the spots enough to determine that they came from an agent small enough to pass through a porcelain filter.

He dropped the investigation, but recalled these observations while studying dysentery in 1915. The illness was affecting soldiers fighting in World War I and he soon determined the cause to be the bacterium *Shigella dysentery.* In the process of investigating *Shigella,* similar areas of clearing were observed. d'Herelle realized that something in these areas of clearing was killing the bacterium. He eventually determined these were viruses of bacteria, coining the term bacteriophage (devourer of bacteria).

THE POWER OF VACCINATION IS DISCOVERED

As soon as it was realized that microbes cause illness, the search was on for ways to kill or to prevent them from causing disease. We will look at two series of events that illustrate the emergence of modern treatment of infectious disease: the development of vaccines and the discovery of antimicrobial compounds.

Smallpox was a feared disease throughout human history and justifiably so. It was highly contagious and almost everyone eventually became infected. Mortality rates were as high as 25 per cent in adults and closer to 40 per cent in children. Those who did survive often had scarring due to the blister-like pustules that form on the skin, but they obtained life-long immunity to the disease.

As far back at the 11th century in India and China it was realized that liquid from the pustules of a smallpox victim, when scratched on the skin of a healthy patient, would most often cause mild disease.

This intentional infection, termed variolation, would also give life-long protection against the virus. Lady Mary Wortley Montgue, wife of ambassador to the Ottoman Empire, introduced variolation to England in 1721 and it became a popular practice throughout Europe. Washington even began variolating the Continental Army in 1776.

Variolation had some deleterious side effects. Serious skin lesions inevitably resulted at the site of inoculation, often accompanied by a generalized rash or even a full case of smallpox. The fatality rate from variolation was 1 to 2 per cent. Today we would find this level of fatality to be unacceptable, but at the time this risk still represented a significant advance.

In 1796, Edward Jenner, an English country physician, went in search of a more predictable and safer method of protection against the disease.

He noticed that milkmaids rarely contracted smallpox. Further investigation revealed they often contracted cowpox from their charges. Jenner hypothesized that cowpox was related to smallpox and contraction of the former would protect against the latter.

In a classic experiment (and one that would land us in jail today), Jenner inoculated a young patient with cowpox and later challenged him with smallpox. The boy did not become ill and Jenner was responsible for the creation of a safer method of protection against smallpox. It is important to stress that the nature of these diseases and their viruses would not be known for over 100 years. Jenner was ahead of his time.

Beginning around 1876 Pasteur's studies on chicken cholera led to the development of vaccines to fight the disease. Cholera was a serious problem since it was able to spread through a barnyard and wipe out a flock in as little as 3 days. It was transmitted by contaminated food or animal excrement. Pasteur identified the cholera bacillus and grew it in pure culture. When injected with it, a chicken invariably died within 48 hours.

Then, as often happens in scientific research, luck intervened. During the heat of the summer, Pasteur returned to Paris and left the cholera cultures used for infection stored on the shelves of his laboratory in Arbois, France. Upon returning, something had happened to the cultures, they no longer caused disease when tested in chickens.

With some impatience for the time they were wasting, his group set to work making new cultures of the bacillus and tested these batches on both new birds and also those previously inoculated with the ineffective strain. To their amazement the previously injected birds were unaffected by the fresh bacillus culture, while the new birds all died. Pasteur immediately realized that this was similar to the studies of Jenner. Pasteur then developed a method for creating cultures that would confer immunity, but not cause disease.

Sometimes this involved growing the microbe in medium in the laboratory where they would spontaneously loose their virulence. In other cases it involved multiple passes through a susceptible host. For example, the rabies virus was attenuated by passing it through rabbits. In honour of Jenner's accomplishments Pasteur coined the term vaccination (*vacca* = cow in Latin) for the process of immunization against disease. In this and several of Pasteur's other discoveries, luck played a part, but it was only helpful because he tenaciously pursued "odd" results and had the insight to arrive at important conclusions. In Pasteur's famous words, "In the field of observation, chance favors only the prepared mind."

Pasteur's technique of weakening a strain by a damaging treatment or passing it through a susceptible host was termed attenuation and resulted in the creation of vaccines against anthrax, plague, yellow fever, rabies and many other diseases. Many vaccines have been developed over the years and children today receive a number of shots, greatly decreasing infant mortality.

ANTIMICROBIAL COMPOUNDS

We now pick up another thread though the web of change that began with the work of Paul Ehrlich. By 1885, it was becoming clear that the causative agents of many illnesses were microorganisms. As scientists manipulated these microbes in the lab they found that certain dyes and other compounds were able to inhibit their growth. This inspired Paul Ehrlich to propose that chemicals may exist that will kill the microbe, but not the patient. The hope was that the chemicals will cure the illness of a patient.

One of the diseases Ehrlich hoped to cure was syphilis, which had reached epidemic proportions in Europe. Little did he realise it would be a 17-year odyssey before he would develop salvarsan, the first effective chemotherapeutic agent. Salvarsan was the 606th chemical he tried. This arsenic compound effectively kills *Treponema pallidum*, the causative agent of syphilis.

The treatment, however, had many problems, causing long lasting health complications for those individuals who used it. In addition, despite *Treponema* being quite sensitive to salvarsan, the physician had to administer it intravenously for optimum effectiveness. Intravenous injection was a recent development and many doctors were leery of trying the procedure. In London, a young physician by the name of Alexander Fleming, then in the Army Medical Corps, was one of the few that was willing to treat patients. Fleming even got the nickname private 606 from his burgeoning practice. His work validated the effectiveness of salvarsan against syphilis and convinced others to administer the treatment.

Fleming, was a physician by training, but spent most of his time studying bacteria, and his success with salvarsan motivated him to search for other antibacterial agents. His first discovery was lysozyme, an enzyme produced by many organisms, including humans that lysed some bacteria.

This enzyme is not useful as a therapeutic agent because it is difficult to administer as a drug, but Fleming did develop titration methods and assays that would become very useful.

Fleming's arguably most important contribution to science is his discovery of penicillin. In September of 1928, before leaving on a summer holiday, Fleming streaked some plates of *Staphylococcus aureus* and left them to incubate until his return. In an improbable set of circumstances, the beginning of the holiday was cold, allowing some contaminating mold spores (that had blown in from a nearby window) to grow up on some of the plates. The temperature then increased encouraging the growth of the *Staphylococcus*.

Many experimenters when confronted with a contaminated plate will look for the trash bin, but Fleming instead spent some time examining it. The fungus had a zone of clearing around it where the *Staphylococcus* colonies would not grow, suggesting the fungus was producing an antibacterial compound that had diffused into the medium. Intrigued, he cultured the fungus, a *Penicillium* mold, and eventually isolated a soluble extract that could kill bacteria and treat localized infection.

He called the new compound penicillin after the mold from which it came. Due to the technology then available, however, it was very difficult to prepare a solution that could be used throughout the body without causing problems.

World War II added a greater urgency to the search for compounds that could fight infectious disease. Wounded soldiers, if they survived the initial injury, would often develop life-threatening infections and there were no effective drugs to combat them. In 1939 Howard Florey and Ernst Chain began a systematic study of antimicrobial compounds in hopes of developing treatments for these soldiers and ran across Fleming's report written 9 years earlier.

They now were able to purify the compound completely and describe its high potency against microbes. The availability of penicillin during World War II saved countless lives. The rediscovery of penicillin touched off a search for other microbes producing substances that could kill or inhibit microbes, leading to the discovery of many more antimicrobials. With Florey and Chain, Fleming was awarded the Noble prize in Medicine and Physiology in 1945.

In the ensuing decades the search and discovery of numerous antimicrobial compounds, combined with the development of vaccines, has eliminated many of the deadly diseases that plagued humankind.

While we now realise this is a continuing war and not a one-time battle, our understanding of these microbes and the nature of disease will likely keep infectious disease at bay for the foreseeable future.

FIELD OF ENVIRONMENTAL MICROBIOLOGY

Contrary to what we would believe from news and advertising, the vast majority of bacteria are harmless to humans and a large number are actually helpful. Understandably, early work in microbiology focused on pathogenic microbes and treatment of the diseases they caused. Martinus Beijerinck and Sergei Winogradsky began the transition from this early work into the present era where we study the molecular biology of a wide variety of microorganisms. Both were interested in microbes present in the soil and water and their work founded the discipline of environmental microbiology.

Much of what we understand about bacteria in the environment and their impact can be traced back to the efforts of these two scientists.

Martinus Beijerinck, was originally trained as a botanist and began his work studying the microorganisms that were present in and around plants. He soon began experiments with microbes in the soil. His greatest contribution was the development of enrichment media. Previously, microbes were cultivated on medium consisting of potatoes or extracts of left-over animal renderings.

Such media would support the growth of many different bacteria, with chance and population density dictating what became dominant in the culture. In many cases this was exactly what was desired, but in other cases it was of interest to find bacteria capable of performing certain chemical conversions.

Beijerinck discovered that by adding or removing certain compounds from the medium or incubating under different conditions, it was possible to favour the growth of certain microbes and prevent the growth of others. An example is Beijerinck's work with nitrogen-fixing bacteria, which are important in agriculture and the global cycling of nitrogen.

They are capable of taking nitrogen gas (N_2) from the air and reducing it to ammonia (NH_3). This is an important property since most organisms, including agriculturally important plants, can only use reduced nitrogen compounds such as ammonia.

Beijerinck wanted to isolate an organism capable of fixing nitrogen in the presence of air, because all previous isolates fixed nitrogen only under anaerobic conditions. By making medium that did not contain a source of fixed nitrogen and then incubating in the presence of air (containing N_2), he demanded that any microbe growing in the medium had to be able to derive its nitrogen by performing aerobic nitrogen fixation. Using this method he succeeded in isolating a new microbe (*Azotobacter chroococum*) with these capabilities.

Similar selective culture techniques (as he liked to call them) enabled his laboratory to isolate sulfur-reducing and sulfur-oxidizing bacteria, *Lactobacillus* species, green algae and many other microbes. Selective culture techniques have since been applied to many different groups of microorganisms, allowing them to easily be brought into pure culture.

Sergei Winogradsky, was also interested in soil bacteria, especially those involved in the cycling of nitrogen and sulfur compounds. He was one of the first to isolate microorganisms responsible for the conversion of these elements in the soil, obtaining pure cultures of bacteria capable of the conversion of ammonia to nitrate by microorganisms in the soil. Winogradsky also studied the consumption of hydrogen sulfide gas by sulfur-oxidizing bacteria directly in their natural habitat.

When working with *Beggiatoa* (one of these sulfur-oxidizing bacteria) he discovered it was obtaining its cell carbon from carbon dioxide. He used the term autotrophy to describe this property, a radical idea at the time, and one not readily accepted by other microbiologists. This led Winogradsky to propose the concept of chemolithotrophy; bacteria capable of growth using purely inorganic sources of carbon and energy. He realized that these microbes were essential for the cycling of nutrients in the environment, and the living world was dependent upon their action, long before others accepted the idea.

As a result of the work of Winogradsky and Beijerinck there was great enthusiasm for identifying and classifying the bacteria inhabiting our natural world. For the first part of the 20th century many scientists isolated microbes and made proposals for their organization into genera and species.

In 1909 Sigurd Orla-Jensen suggested a classification scheme based on the functions and abilities of the bacteria, such as growth on certain compounds or the production of specific by-products. The Society of American Bacteriologists, later to become the American Society of Microbiology, applied this technique to prepare a report on the classification of bacteria. This eventually evolved into Bergey's Manual of Determinative Bacteriology in 1923 and subsequent editions have become authoritative reference works.

There was great optimism that as more bacteria were brought into pure culture a clear organization, based on the evolutionary history of microbes, would emerge from studies of their physiology. However, by the 1940's it became clear that bacteria were unwilling to go along with this idea because classification based on one set of properties was often inconsistent with classification using a different set of properties. Scientists threw up their hands and gave up the idea of finding a system of classification based on growth and morphological characteristics.

Several decades passed during which bacteria became tools for understanding life, but their place in evolution was of little interest. Then in 1961 Brian McCarthy and E. T. Bolton developed methods for the comparison of genetic material between species and Linus Pauling and Emile Zuckerkandl formalized the idea of using the makeup of biological molecules, their sequence, as a way to determine phylogenetic relationships.

Later, Carl Woese, began to use this insight, choosing the sequence of the 16S rRNA from the ribosomes of bacteria as basis for evolutionary comparisons. This analysis was a watershed event for the evolutionary classification of bacteria and in 1977 Woese used this technique to assert that known bacteria should be divided into two separate domains that we now call Bacteria and Archaea, radically changing our view of the microbial world.

LIFE AT THE MOLECULAR LEVEL

Microbes have been critical tools for the understanding of basic biology at the molecular level. Fred Griffith performed transformation experiments that suggested that DNA was the hereditary material. His student, Avery Oswald, later demonstrated this conclusively. Beadle and Tatum developed the one gene-one enzyme hypothesis using *Neuropsora crassa* as the test microbe.

Many scientists working with *E. coli* deciphered the nature of DNA, RNA and protein. These experiments lead to our understanding of how information in the cell is stored and how it is converted into proteins.

In the early part of the 20th century techniques were developed to examine the inner workings of the cell and much of the work was performed in bacteria due to their experimental accessibility. Before this period, the method of turning genetic information into the proteins that carry out cellular processes was completely unknown. Indeed, the chemical composition of the genetic material was being hotly debated. From Gregor Mendel's work with pea plants, the nature of inheritance and heredity was understood. However, it was unclear what molecule in the cell maintained and passed hereditary information on to subsequent generations. The leading contender for this role was protein, when in 1928 Fred Griffith discovered transformation in bacteria.

Griffith knew that when *Streptococcus pneumoniae* strains are injected into mice, they cause rapid deterioration and death. Griffith isolated strains of the bacteria that were no longer capable of producing an outer slime layer and appeared rough when grown on solid medium in contrast to the smooth colonies of the original isolate.

When these rough strains were injected into mice, no illness resulted. Similarly, when a smooth microbe was heat-killed and then injected, the mice showed no signs of infection. A surprise was waiting for Griffith when he injected into mice a mixture of dead cells of a smooth isolate with live cells of a rough isolate: they died. When bacteria were isolated from these dead mice, they formed smooth colonies. Griffith hypothesized that the ability to create the slime layer was passed from the dead smooth cells to the viable rough cells, making them pathogenic again. The process became known as transformation.

Griffith was ridiculed, as most scientists believed his preparation were contaminated with viable smooth cells. It was not until 1944 that his student Oswald Avery and coworkers repeated the experiments of Griffith, reproducing his results and discovering that DNA was the material from the dead smooth cells that transformed the rough mutant. This was strong proof that the hereditary material in cells was DNA. Further experiments in the 1940's clearly established this observation with Joshua Lederberg discovering two more ways that DNA could be transferred between bacterial cells, conjugation and transduction.

In 1943 Beadle and Tatum reported experiments with the fungus *Neuropsora crassa* that eventually established the idea that each gene in the DNA typically codes for one protein (the one gene-one enzyme hypothesis). About 10 years later, Francis Crick, Rosalind Franklin, James Watson and Maurice Wilkins worked on experiments describing the structure of DNA and making predictions about how it was replicated. It was now clear that DNA stored the information for proteins and that proteins performed the many functions of the cell.

The important question now became, how does one convert the information in DNA into protein. A major contribution in understanding this puzzle was made by Paul Zemecnik and his lab who developed cell-free systems, first with rat liver and later using the bacterium *E. coli*. This allowed the study of translation in the test tube and he and other scientists used these systems to discover the important molecules involved in the process.

An intense effort to describe these molecular events then ensued. The first light came when Crick, Sydney Brenner and colleagues proposed the existence of transfer RNA (tRNA), a molecule that helps to create amino acid polymers based on nucleic acid sequence.

Another critical player in the processing of information was revealed when Brenner, Francois Jacob and Matthew Meselson discovered that translation of genetic material into protein takes place on the ribosome and that the molecule being translated at the ribosome is RNA, not DNA. The next mystery was solved by Marshal Niremberg and J. H. Matthaei when they developed methods to decipher the genetic code that dictates the correspondence of nucleic acids to amino acids. After the stunningly productive decade of the 1960's, the nature of the framework for the conversion of the genetic information into proteins was now understood and the basic mechanism has since been shown to be conserved across all biology.

TAKING CONTROL OF THE GENOME

One final thread we will trace is the rise of molecular

microbiology and the technique of genomics. This pathway begins in 1885 with the seemingly unimportant isolation of a common intestinal microorganism by Theodor Escherich.

The microbe is eventually renamed *Escherichia coli* in his honour. It could not have been understood at the time, but this bacterium would play a central role in our understanding of life. *E. coli* achieved this position because it has several properties that made it invaluable as an experimental system:

- *E. coli* is easy to take care of. It can grow on a wide variety of media, at temperatures ranging from 20-45 °C, in the presence or absence of air.
- *E. coli* grows quickly. Under normal laboratory conditions it will go through a complete life cycle, doubling its population about every 30 minutes. With this rapid growth rate it is possible to generate large populations in just a few hours. Most experiments can be set up one day and the results observed the next.
- It can survive storage for long periods of time without losing viability using relatively simple preservation techniques.
- It was available. There is an element of arbitrariness that this microbe came to serve such a vital role in biology. Many other bacteria probably could have filled this position, but *E. coli* was the one chosen.

Several key discoveries laid the foundation for future developments in molecular biology. The first of these was the discovery of autonomously replicated pieces of DNA separate from the bacterial chromosome. Plasmid as Joshua Lederberg called them, were capable of carrying genes that could be regulated and moved from one microbe to another independently of the bacterial chromosome.

The ability of some plasmids to move between different cells meant that different bacteria could rapidly inherit genes carried on these plasmids.

It was found that one class of plasmids encoded resistance genes that would enable the bacteria to grow in the presence of an antibiotic. It later became clear it was possible to select for the presence of these plasmids by demanding growth of the bacterial strain on medium containing the antibiotic to which the plasmid encoded resistance.

A second discovery that greatly impacted modern genetic engineering was that bacteria have their own parasites. As mentioned earlier, there are viruses that infect and kill bacteria, just as viruses infect animals and plants. The analysis of bacterial viruses was a major avenue to understanding the bacteria themselves.

For example, investigation of the ability of some viruses to grow well on certain strains of bacteria and not others lead to the discovery of restriction-modification systems in bacteria. These systems allow bacteria to recognize and destroy foreign DNA and consist of two parts.

A modification enzyme that labels DNA as belonging to the bacterium and a restriction enzyme that recognizes a certain 4-8 base pair sequence in the DNA and cuts it in two if it has not been modified. The landmark paper of Hamilton Smith and Kent Wilcox showed that restriction enzymes were capable of cleaving double-stranded DNA into discrete pieces. The authors correctly hypothesized that the enzyme was recognizing a sequence of DNA and cutting it.

Further studies on restriction enzymes led to the birth of genetic engineering. Joan Mertz and Ronald Davis made restriction enzymes into tools for the manipulation of DNA when they showed that many of the breaks these enzymes made in the DNA produced single-stranded complementary ends. Stanley Cohen, Annie Chang, Robert Helling and Herbert Boyer then demonstrated that any DNA can be broken into fragments with restriction enzymes and by mixing that with a plasmid digested in a similar manner, it was possible to create recombinant molecules.

Moving the plasmid into a microbe and growing it on selective medium could produce any desired amount of these recombinant molecules.

Another discovery was "brewing" in the western United States. In the mid 1960s, Thomas Brock was on a trip to Yellowstone and became intrigued by the mats of green, brown and pink material that were found in many of the hot springs.

Brock was sure that these were living communities of microorganisms, yet the springs were at temperatures near boiling. Subsequent researches proved his hypothesis correct and lead to the isolation of a microbe, *Thermus aquaticus,* capable of growth at 85 °C.

Organisms capable of growing at this high a temperature were unheard of at the time and the discovery spawned two major developments. First, work with *T. aquaticus* and other unusual microbes by Woese and many others lead to the discovery of the Archaea. Second, the enzyme responsible for replicating DNA in *T. aquaticus* was found to be extremely heat stable, surviving temperatures of 95°C and copying DNA at 72°C. The use of *T. aquaticus* polymerase sped the adoption of polymerase chain reaction (always referred to as PCR).

This technique was developed in 1985 by Kary Mullis, and became incredibly useful for amplifying even small quantities of DNA. Originally the technique required fresh addition of polymerase enzyme after each step, which could become prohibitively expensive. The use of *T. aquaticus* polymerase made the procedure much more economical. PCR has become a pivotal technique in many areas of science, from detecting bacteria in the environment to the analysis of evidence at crime scenes.

In 1977 Walter Gilbert and Fred Sanger independently developed methods for determining the exact sequence of bases in DNA. These techniques became immediately useful in determining the sequence of numerous important genes being investigated in countless laboratories.

Further refinement has made DNA sequencing more efficient. It became efficient enough that in 1985 Robert Sinsheimer convened a meeting of a number of biologists active in genetics and gene mapping. Out of this meeting came a proposal to sequence the entire three billion base pairs of the human genome.

This was equivalent at the time of proposing a mission to the moon, technically possible, but with unpredictable value. The initial cost of the project was estimated to be $10 a base pair or a staggering 30 billion dollars. There was resistance from a significant portion of the scientific community, fearing that the project would siphon funds away from other worthy scientific projects. However, the clarity of the goal ignited the imagination of Congress and the public, creating unstoppable momentum for the project.

After about a decade of effort, a preliminary draft of the entire genome was released in the spring of 2001, earlier than projected and dramatically under budget. The project itself drove the development of sequencing technology, allowing the determination of millions of base pairs of sequence per day at a cost of less than 3 cents per base at large sequencing facilities.

Along with the human genome, many other model organisms have been or are about to be sequenced. These include the genomes of many pathogenic bacteria, the mouse, the fruit fly and the nematode *Caenorhabditis elegans,* the last being an experimental model for development in eukaryotes.

Microbiology is the study of organisms that at some point in their life exist as single cells and contain a nucleic acid genome that can replicate. Many organisms fall into this definition including algae, fungi, protozoa, bacteria and archaea. Together these organisms have a profound impact on the biosphere, making up the majority of life both in number and total mass. Many illnesses are caused by infection with microbes and understanding these infections has lead to cures and better treatments.

The emergence of new infectious agents will spur continued interest in microbiology. Many more microbes grow harmlessly in the environment, taking advantage of chemicals and/or sunlight to grow. Research into these microbes has also helped us understand the basic framework of life and revealed the basic fundamental rules that govern living systems. In the past microbes have been used in experiments to answer many scientific questions and they will continue to serve as excellent tools of inquiry in the future. A significant number of these discoveries have lead to important applications in many areas of human endeavor.

Chapter 8

Practical Utilisation of Microorganisms

MICROBIAL BIODEGRADATION

Interest in the microbial biodegradation of pollutants has intensified in recent years as mankind strives to find sustainable ways to cleanup contaminated environments. These bioremediation and biotransformation methods endeavour to harness the astonishing, naturally occurring, ability of microbial xenobiotic metabolism to degrade, transform or accumulate a huge range of compounds including hydrocarbons (e.g. oil), polychlorinated biphenyls (PCBs), polyaromatic hydrocarbons (PAHs), pharmaceutical substances, radionuclides and metals.

Major methodological breakthroughs in recent years have enabled detailed genomic, metagenomic, proteomic, bioinformatic and other high-throughput analyses of environmentally relevant microorganisms providing unprecedented insights into key biodegradative pathways and the ability of organisms to adapt to changing environmental conditions.

The elimination of a wide range of pollutants and wastes from the environment is an absolute requirement to promote a sustainable development of our society with low environmental impact.

Biological processes play a major role in the removal of contaminants and they take advantage of the astonishing catabolic versatility of microorganisms to degrade/convert such compounds. New methodological breakthroughs in sequencing, genomics, proteomics, bioinformatics and imaging are producing vast amounts of information. In the field of Environmental Microbiology, genome-based global studies open a new era providing unprecedented *in silico* views of metabolic and regulatory networks, as well as clues to the evolution of degradation pathways and to the molecular adaptation strategies to changing environmental conditions.

Functional genomic and metagenomic approaches are increasing our understanding of the relative importance of different pathways and regulatory networks to carbon flux in particular environments and for particular compounds and they will certainly accelerate the development of bioremediation technologies and biotransformation processes.

Aerobic Biodegradation of Pollutants

The burgeoning amount of bacterial genomic data provides unparalleled opportunities for understanding the genetic and molecular bases of the degradation of organic pollutants. Aromatic compounds are among the most recalcitrant of these pollutants and lessons can be learned from the recent genomic studies of *Burkholderia xenovorans* LB400 and *Rhodococcus* sp. strain RHA1, two of the largest bacterial genomes completely sequenced to date.

These studies have helped expand our understanding of bacterial catabolism, non-catabolic physiological adaptation to organic compounds, and the evolution of large bacterial genomes. First, the metabolic pathways from phylogenetically diverse isolates are very similar with respect to overall organization.

Thus, as originally noted in pseudomonads, a large number of "peripheral aromatic" pathways funnel a range of natural and xenobiotic compounds into a restricted number of "central aromatic" pathways.

Nevertheless, these pathways are genetically organized in genus-specific fashions, as exemplified by the b-ketoadipate and Paa pathways. Comparative genomic studies further reveal that some pathways are more widespread than initially thought. Thus, the Box and Paa pathways illustrate the prevalence of non-oxygenolytic ring-cleavage strategies in aerobic aromatic degradation processes. Functional genomic studies have been useful in establishing that even organisms harboring high numbers of homologous enzymes seem to contain few examples of true redundancy.

For example, the multiplicity of ring-cleaving dioxygenases in certain rhodococcal isolates may be attributed to the cryptic aromatic catabolism of different terpenoids and steroids. Finally, analyses have indicated that recent genetic flux appears to have played a more significant role in the evolution of some large genomes, such as LB400's, than others.

However, the emerging trend is that the large gene repertoires of potent pollutant degraders such as LB400 and RHA1 have evolved principally through more ancient processes. That this is true in such phylogenetically diverse species is remarkable and further suggests the ancient origin of this catabolic capacity.

Anaerobic Biodegradation of Pollutants

Anaerobic microbial mineralization of recalcitrant organic pollutants is of great environmental significance and involves intriguing novel biochemical reactions.

In particular, hydrocarbons and halogenated compounds have long been doubted to be degradable in the absence of oxygen, but the isolation of hitherto unknown anaerobic hydrocarbon-degrading and reductively dehalogenating bacteria during the last decades provided ultimate proof for these processes in nature. Many novel biochemical reactions were discovered enabling the respective metabolic pathways, but progress in the molecular understanding of these bacteria was rather slow, since genetic systems are not readily applicable for most of them.

However, with the increasing application of genomics in the field of environmental microbiology, a new and promising perspective is now at hand to obtain molecular insights into these new metabolic properties. Several complete genome sequences were determined during the last few years from bacteria capable of anaerobic organic pollutant degradation. The 4.7 Mb genome of the facultative denitrifying *Aromatoleum aromaticum* strain EbN1 was the first to be determined for an anaerobic hydrocarbon degrader (using toluene or ethylbenzene as substrates).

The genome sequence revealed about two dozen gene clusters (including several paralogs) coding for a complex catabolic network for anaerobic and aerobic degradation of aromatic compounds. The genome sequence forms the basis for current detailed studies on regulation of pathways and enzyme structures.

Further genomes of anaerobic hydrocarbon degrading bacteria were recently completed for the iron-reducing species *Geobacter metallireducens* and the perchlorate-reducing *Dechloromonas aromatica,* but these are not yet evaluated in formal publications. Complete genomes were also determined for bacteria capable of anaerobic degradation of halogenated hydrocarbons by halorespiration: the 1.4 Mb genomes of *Dehalococcoides ethenogenes* strain 195 and *Dehalococcoides* sp. strain CBDB1 and the 5.7 Mb genome of *Desulfitobacterium hafniense* strain Y51.

Characteristic for all these bacteria is the presence of multiple paralogous genes for reductive dehalogenases, implicating a wider dehalogenating spectrum of the organisms than previously known. Moreover, genome sequences provided unprecedented insights into the evolution of reductive dehalogenation and differing strategies for niche adaptation.

Bioavailability, Chemotaxis, and Transport of Pollutants

Bioavailability, or the amount of a substance that is physiochemically accessible to microorganisms is a key factor in the efficient biodegradation of pollutants. Chemotaxis, or the directed movement of motile organisms towards or away from chemicals in the environment is an important physiological response that may contribute to effective catabolism of molecules in the environment.

In addition, mechanisms for the intracellular

accumulation of aromatic molecules via various transport mechanisms are also important.

Oil Biodegradation

Petroleum oil contains aromatic compounds that are toxic for most life forms. Episodic and chronic pollution of the environment by oil causes major ecological perturbations. Marine environments are especially vulnerable since oil spills of coastal regions and the open sea are poorly containable and mitigation is difficult. In addition to pollution through human activities, millions of tons of petroleum enter the marine environment every year from natural seepages.

Despite its toxicity, a considerable fraction of petroleum oil entering marine systems is eliminated by the hydrocarbon-degrading activities of microbial communities, in particular by a remarkable recently discovered group of specialists, the so-called hydrocarbonoclastic bacteria (HCB). *Alcanivorax borkumensis* was the first HCB to have its genome sequenced.

ANALYSIS OF WASTE BIOTREATMENT

Sustainable development requires the promotion of environmental management and a constant search for new technologies to treat vast quantities of wastes generated by increasing anthropogenic activities. Biotreatment, the processing of wastes using living organisms, is an environmentally friendly, relatively simple and cost-effective alternative to physico-chemical clean-up options. Confined environments, such as bioreactors], have been engineered to overcome the physical, chemical and biological limiting factors of biotreatment processes in highly controlled systems.

The great versatility in the design of confined environments allows the treatment of a wide range of wastes under optimized conditions. To perform a correct assessment, it is necessary to consider various microorganisms having a variety of genomes and expressed transcripts and proteins. A great number of analyses are often required. Using traditional genomic techniques, such assessments are limited and time-consuming. However, several high-throughput techniques originally developed for medical studies can be applied to assess biotreatment in confined environments.

Metabolic Engineering and Biocatalytic Applications

The study of the fate of persistent organic chemicals in the environment has revealed a large reservoir of enzymatic reactions with a large potential in preparative organic synthesis, which has already been exploited for a number of oxygenases on pilot and even on industrial scale.

Novel catalysts can be obtained from metagenomic libraries and DNA sequence based approaches. Our increasing capabilities in adapting the catalysts to specific reactions and process requirements by rational and random mutagenesis broadens the scope for application in the fine chemical industry, but also in the field of biodegradation. In many cases, these catalysts need to be exploited in whole cell bioconversions or in fermentations, calling for system-wide approaches to understanding strain physiology and metabolism and rational approaches to the engineering of whole cells as they are increasingly put forward in the area of systems biotechnology and synthetic biology.

XENOBIOTIC

A xenobiotic is a chemical which is found in an organism but which is not normally produced or expected to be present in it. It can also cover substances which are present in much higher concentrations than are usual. Specifically, drugs such as antibiotics are xenobiotics in humans because the human body does not produce them itself nor would they be expected to be present as part of a normal diet. However, the term is also used in the context of pollutants such as dioxins and polychlorinated biphenyls and their effect on the biota. Natural compounds can also become xenobiotics if they are taken up by another organism (e.g., uptake of natural human hormones by fish found downstream of sewage treatment plant outfalls).

Xenobiotic Metabolism

The body removes xenobiotics by xenobiotic metabolism.

This consists of the deactivation and the secretion of xenobiotics, and happens mostly in the liver.

Secretion routes are urine, feces, breath, and sweat. Hepatic enzymes are responsible for the metabolism of xenobiotics by first activating them (oxidation, reduction, hydrolysis and/or hydration of the xenobiotic), and then conjugating the active secondary metabolite with glucuronic or sulphuric acid, or glutathione, followed by excretion in bile or urine. An example of a group of enzymes involved in xenobiotic metabolism is hepatic microsomal cytochrome P450. These enzymes that metabolize xenobiotics are very important for the pharmaceutical industry, because they are responsible for the breakdown of medications.

Xenobiotics in the Environment

Xenobiotic substances are becoming an increasingly large problem in Sewage Treatment systems, since they are relatively new substances and are very difficult to categorize. Antibiotics, for example, were derived from plants originally, and so mimic naturally occurring substances. This, along with the natural monopoly nature of municipal Waste Water Treatment Plants makes it nearly impossible to remove this new pollutant load.

Some xenobiotics are resistant to degradation. For example, they may be synthetic organochlorides such as plastics and pesticides, or naturally occurring organic chemicals such as polyaromatic hydrocarbons (PAHs) and some fractions of crude oil and coal. However, it is believed that microorganisms are capable of degrading all the different complex and resistant xenobiotics found on the earth.

Inter-Species Organ Transplantation

The term xenobiotic is also used to refer to organs transplanted from one species to another. For example, some researchers hope that hearts and other organs could be transplanted from pigs to humans. Many people die every year whose lives could have been saved if a critical organ had been available for transplant. Kidneys are currently the most commonly transplanted organ. Xenobiotic organs would need to be developed in such a way that they would not be rejected by the immune system. With the development of vitrification transplantable organs could be stored in organ banks for long periods.

MICROBIAL DEGRADATION OF XENOBIOTICS

Microorganisms play an essential role in the bioconversion and total breakdown of pesticides and other xenobiotics in the environment. In recent years,a number of problems related to microbial degradation of pesticides have been encountered. Some examples are given below. In certain 'problem' soils, the efficacy of applied pesticides is reduced due to their enhanced rate of degradation.In other cases, environmental accumulation of persistent pesticides has become a problem. Some pesticide residues are bound by the soil but under specific conditions may be released due to microbial attack.Most bioconversions are detoxification reactions.

However, there have been reports of the formation of more toxic pesticidal metabolites. Certain pesticides are leached down to deeper subsoil horizons or underground water systems but, to date, our understanding of microbial processes occurring in these systems and undoubtedly affecting the fate of pesticides is far from adequate. In addition, there is limited knowledge of the conversion of pesticides in other anaerobic zones, e.g. water sediments.

The relevant literature shows that microorganisms possess high potential for bioconversion and total degradation of xenobiotics. This is especially true now that modern genetic engineering enables successful manipulation of specific abilities of microorganisms.

The present document presents a critical review of the mechanisms of microbial bioconversion of xenobiotics, indicates various problems and outlines a basic rational approach to these problems using recent developments in microbiology and biotechnology.

BIOCONVERSION OF XENOBIOTICS

Many xenobiotics contain functional groups and partial

structures not usually found in nature. Many pesticides are bulky complex molecules, and others display hydrophobic properties due to the presence of reduced hydrocarbon fragments. The principal enzymes responsible for the bioconversion of xenobiotics are various lyases and oxydoreductases, specifically hydrolases, oxygenases and various enzymes capable of dehalogenation.

Hydrolysis

Amide and ester bonds undergo hydrolytic cleavage in anilides, phenylureas, esters of carbamic, thiocarbamic, phosphoric, thiophosphoric, and of other acids. Hydrolases responsible for the cleavage of pesticides are among the best studied groups of enzymes. Most of these hydrolases are extracellular enzymes, except for the cell wall-bound enzymes of Penicillium and Arthrobacter sp., which hydrolyse barban and propham, respectively. Aryl- and acylamidases, which hydrolyse the amide bond in phenylamides, have been studied and described in sufficient detail. Usually, these enzymes are induced by a broad spectrum of substrates and exhibit low substrate specificity.

Table: Induction of acylamidase in Bacillus sphaericus 12123

Inductor 10^{-3} unit/mg of linuron	*Specific activity per cent of that*	*Induction rate, protein*
Linuron	0.13	100
Maloran	0.015	12
Monalide	0.029	22
Propanil	0.002	2
Propham	0.008	6
2–Chlorobenzanilide	0.03	23
2, 5–Dimethylfuran–3–Carboxanilide	0.05	40

In 1976, Munnecke and Hsieh reported similar results: a culture capable of the thioester bond cleavage (416 nmoljmin per 1 mg protein) also brought about enzymic hydrolysis of another 8 of the 12 insecticides tested.

This ability of microbial hydrolases underlies the

emergence of 'problem' soils, which are capable of hydrolyzing a wide variety of pesticides.

Commission on Agrochemicals

Table: Hydrolysis of Phenylamides by a cell-free extract of Bacillus Sphaericus induced by linuron

Phenylamide	*Specific activity 10^{-3} unit/mg protein*
Linuron	0.20
Monolinuron	0.20
Maloran	0.11
Carboxin	2.52
2–Chlorobenzanilide	18.7
Glycine 4–Nitroanilide	106.0
L–Alanine 4–Nitroanilide	309.0
L–Leucine 4–Nitroanilide	55.8
L–Phenylalanine 4–Nitroanilide	7.63

Dehalogenation

Many xenobiotics contain halide atoms bonded to aliphatic or aromatic carbon. Therefore, dehalogenation reactions are important initial steps in the degradation of such xenobiotics. Although molecular mechanisms of dehalogenation are not fully understood and require further studies, certain reaction types are fairly well known: hydrolytic, reductive and oxidative dehalogenations.

The latter type includes dehydrohalogenation, i.e. dehalogenation which involves molecular oxygen, and results in the formation of a double bond. A number of enzymes have been isolated, described and characterized as effecting dehalogenation. These are haloacetate-halidohydrolase which splits off a halogen from halide acetate to form glycolic acid; hydrolases effecting dehalogenation of dichloroacetate; 2-haloacid halidohydrolase.

The above enzymes are responsible for the hydrolytic release of a halide from the 2 position of short-chain fatty acids. The enzymes which carry out reductive dehalogenation of aromatic carbon have not yet been characterized, although literature provides a number of examples of the process including with DDT, 2,4-D and polychlorophenols. As far as oxidative dehalogenation is concerned, the enzyme - DDT-dehydrochlorinase has been characterized and descriptions of the dioxygenase-mediated dechlorination of picloram and monooxygenase dechlorination of chlorophenoxyacids, chlorobenzoic acids, and chlorophenols are available.

Oxidation

Oxidative processes are very important in the bioconversion of pesticides and other xenobiotics, especially bioconversions which follow hydrolysis and dehalogenation of the parent. In this case, the basic reactions are hydroxylation, followed by cleavage of the aromatic ring, oxidative 0- and N-dealkylation, and epoxidation. Certain compounds, e.g. derivatives of aromatic acids, are completely degraded due to the involvement of oxidative enzymes.

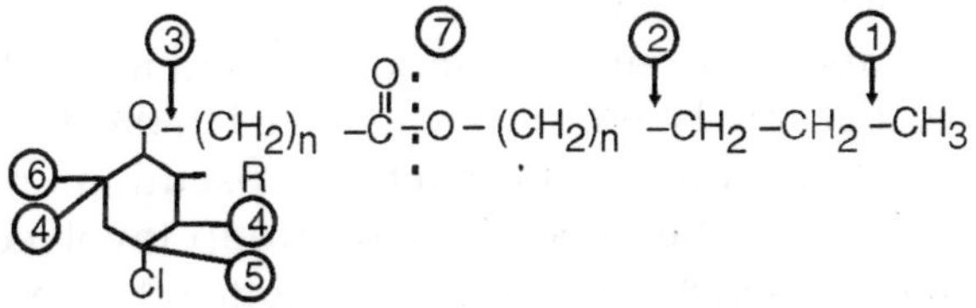

Sites of oxidative attack by enzymes during degradation of alkyl esters of chlorophenoxyacides

R – Methyl or Cl;
1 – Hydroxylation of terminal methyl;
2 – w– oxidation;
3 – Oxidative cleavage of ether bond;
4 – Hydroxylation of the ring;
5 – Oxidative dechlorination;
6 – Cleavage of the ring;
7 – Dashed line – possible hydrolytic fission by esterase.

The above bioconversions are adequately described in Hayaishi's monograph.

Reduction

In general, rezuctive processes take place under anaerobic

conditions in the early stages of degradation. Mention has already been made of reductive dechlorination. Another no less important process is reduction of nitro groups. Enzymes which bring about this process are called nitroreductases.

They perform their functions in the presence of NADH, the reaction being stimulated by the reduced form of FAD, as well as by Mn and Fe ions. A number of authors reported that these enzymes have low substrate specificity. Thus, the enzymic preparation from Vellonella alkalescens was found to catalyse the reduction of nitro groups of some 40 different compounds. The amino products of nitro group reduction may resist further transformation due to the formation of bound residues.

MICROORGANISMS EFFECTING DETOXIFICATION

A review of the research on microbial degradation of xenobiotics reveals that bacteria are chiefly responsible for the detoxification of xenobiotics.

Fungi and yeasts are less important, and microalgae and Protozoa appear to be only rarely involved in the degradation of xenobiotics. Among the bacteria, pseudomonads are considered to be the most efficient group involved in the degradation of xenobiotics. Bacilli are quite often responsible for processes of pesticide hydrolysis. However, pseudomonads can bring about hydrolysis as successfully as they do dehalogenation, hydroxylation, and aromatic ring cleavage and nitro group reduction.

The bacterial genera Acinetobacter, Arthrobacter, Rhodococcus and Flavobacterium also deserve mention, since they often bring about various bioconversions of different xenobiotics. Fungal cultures, especially the genera Aspergillus and Penicillium, are often involved in bioconversions of 8-triazines. Certain fungi have also been reported to bring about methylation of hydroxy- and amino groups, and metals.

Thus the fungus Trichoderma virgatum effected the methylation of pentachlorophenol, whereas Penicillium notatum, Aspergillus Niger and Scopulariopsis brevicaulis methylated arsenic derivatives.

Table: Bioconversion of Xenobiotics Effected by Pseudomonads

Mode of action	*Species*
Hydolysis of carbaryl, dichlorphos, parathion	Pseudomonas diazination, Melophthora
Hydrolysis of parathion	Pseudomonas Stutzeri
Dehalogenation of halide acetate	Pseudomonas sp.
Total dehalogenation of DDT, aromatic ring cleavage	Pseudomonas aeruginosa
Total degradation of 3–chlorobenzoate	Pseudomonas putida
Oxidative dehalogenation of lindane	Pseudomonas putida
Reduction of nitro group in 4, 6–dinitro–Q-Cresol	Pseudomonas sp.
Total degradation of 2,4,5–T	Pseudomonas sepacia
Degradation of toluene, xylenes,	Pseudomonas putida
styrene, a–methylstyrene	Pseudomonas aeruginosa

ROLE OF MICROORGANISMS IN DETOXIFICATION OF PESTICIDES

The important role played by microorganisms in the degradation of pesticide residues is well recognized. According to various authors, microorganisms degrade 25 to 70 per cent of the xenobiotic residues. Literature carries reports on microbial strains capable of degrading such recalcitrant pesticides as DDT polychlorobiphenyls, dicofol, 2,4,5-T and others. Unfortunately, very few xenobiotics can be utilized by pure cultures as the sole source of carbon.

Those, which can, are fairly simple compounds such as dalapon, chlorobenzoates, 2,4-D. Of greater significance are bioconversions of pesticides affected by mixed cultures when the initial degradation steps are brought about by one culture, and metabolites produced are further degraded by other microbial strains. One example is the hydrolysis of parathion by Pseudomonas stutzeri to produce nitrophenol which is further degraded by another strain - Pseudomonas aeruginosa. Exchange of genetic information by natural miCrObia1 populations also appear s to be a potent factor facilitating total degradation of xenobiotics.

For example, one hypothesis states that phenoxyalkane carboxyherbicides, such as 2,4-D and MCPA, do not accumulate in the environment exactly because plasmids, which several years ago were found to encode their degradation, can be transferred under natural conditions from strain to strain (ref. 19). Recently, plasmids responsible for biodegradation of polychlorobiphenyls, chlorobenzoates, petroleum, SUK factants and other compounds were discovered. They apparently play an important role in degradation of numerous foreign compounds.

MICROBIAL BIOCONVERSION OF XENOBIOTICS

Most xenobiotics are readily degraded in natural ecosystems. However, some compounds are transformed into intermediates which are highly resistant to microbial attack or are more toxic than the parent compound. The hydrolysis of phenylamide herbicides, for instance, results in the formation of chloroanilines, which under the action of microbial enzymes are condensed to form tetrachlorohydrazobenzenes:

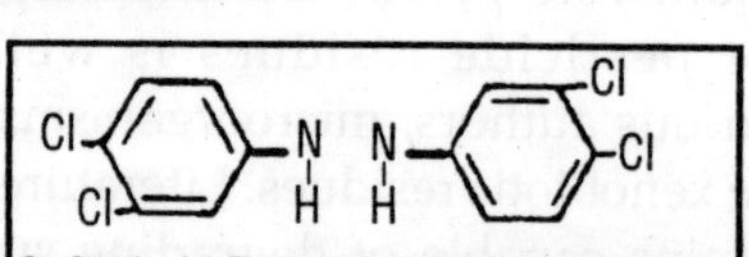

3, 3′, 4, 4′–Tetrachlorohydrazobenzene

As a result, polychloroaromatic compounds appear in the biosphere. These are much more resistant to degradation and more toxic than the parent pesticides and chloroanilines. Thus Kearney and co-workers have detected 3, 3', 4, 4tetrachloro-hydrazobenzene in soil of rice fields some three years after their treatment with propanil. It should be stated that peroxidase and, to a lesser extent, aniline oxidase are responsible for the production of chlor ohydr azobenzenes. Similar processes also occur dux ing the degradation of carbaryl (1-naphthyl N-methylcar bamate), where the hydrolysis of carbaryl leads to á -naphthol which under the effect of phenol oxydase is condensed to give complex compounds highly toxic for aquatic organisms.

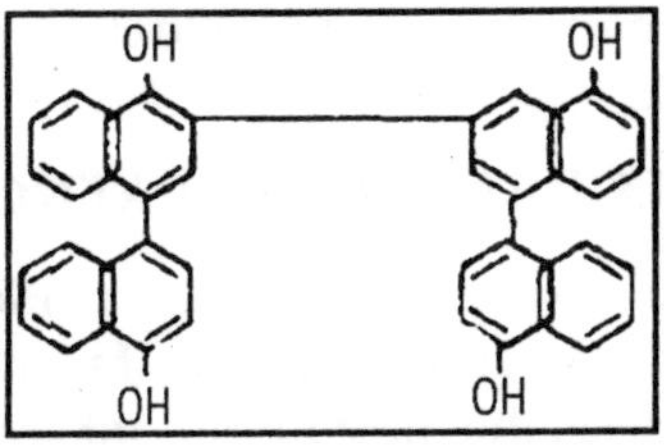

A particular threat for the environment are the metabolites of nitrogen-containing pesticides, which on further degradation are readily transformed in highly mutagenic and carcinogenic compounds. FOL example, simazine is' oxidized into a carcinogen of the following structure.

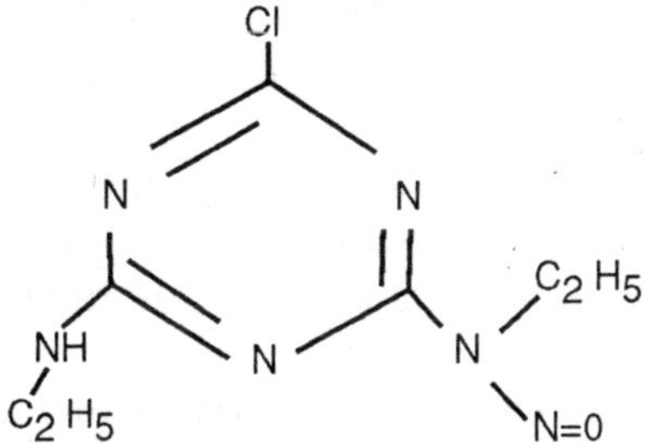

Two of the trifluralin degradation products are also hazardous as mutagenic factors:

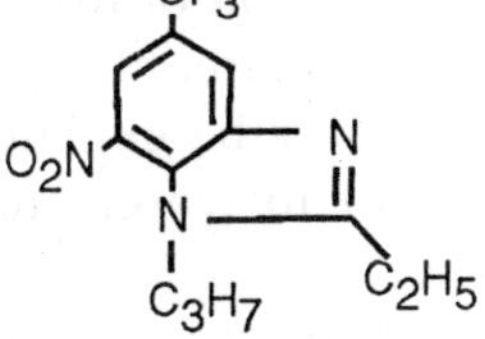

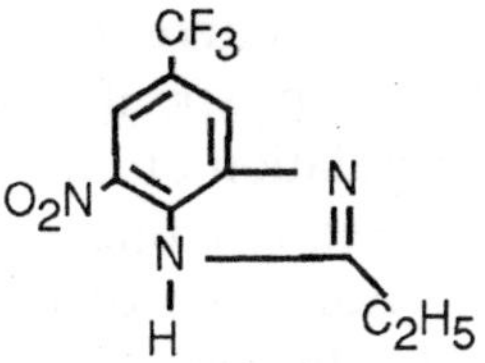

These examples show the need for the thorough studies of the properties of the intermediate products from the microbial bioconversion of pesticides. Their effect on biocenoses and man may be more deleterious than that of parent xenobiotics.

MICROBIAL DEGRADATION OF XENOBIOTICS

The principal cause of pesticide persistence in soil is commonly the lack of conditions favouring microbial degradation. Of these, the most important are:low accessibility of toxicants to enzymes of microorganisms (determined by the extent of the sorption of toxicant by soil particles and colloids), availability of organic compounds utilized by the microflora as energy substrates, moisture conditions, aeration levelltemperature, pH etc.

Certain pesticides are readily degradable in liquid media but, in soil, become resistant to microbial attack. A classical example is the herbicide paraquat which is strongly sorbed by soil thus becoming inaccessible to microorganisms. Quite often the primary transformation of a pesticide, e.9. hydrolysis, leads to phenols or anilines followed by their strong binding to soil.

This results in their inaccessibility to microorganisms. The data concerning the effect of organic matter in soil on microbial degradation of xenobiotics are rather conflicting. Some authors report an inhibiting effect of increasing organic matter on the rate of microbial decomposition of toxicants in soil, whereas others report that organic matter stimulates biodegradation. The contradictory effect may be explained by the dual role played by organic matter in water and soil in relation to microorganisms.

In the latter case, it may provide a source of energy and factors required for the occurrence of primary steps of xenobiotic degradation. Finally, specific components of organic matter may affect the regulation of microbial enzymes.

An explanation of the stimulating action of organic matter on microbial degradation was provided by studies of cometabolism. This is defined as the degradation or bioconversion of organic compounds, mainly of xenobiotics, by microorganisms in the presence of more readily accessible substrates. Biologically, cometabolism provides the processes of xenobiotic metabolism with energy, cofactors, effectors and, possibly, oxidants produced in the course of the cosubstrate metabolism. Cometabolism of pesticides with various cosubstrates has been described for both laboratory and natural conditions.

Thus, the herbicidemolinatewas degraded by cultures of Bacillus sp. in the presence of ethanol: DDT was found to be cometabolized by the strain P.aeruginosa 640 x faster on a medium with glycerol and hexadecane, and alvison during microbial growth on organic acids.

Numerous experiments, on the introduction of cosubstrates into soils and aquatic systems, showed a substantial increase of the soil microflora activity towards simazine, dexon, diazinon, pentachloronitrophenol, paraquat and other pesticides. Hence, the polymeric organic matter of soil colloidal systems may impede microbial degradation of pesticides as a result of their adsorption. In contrast, low molecular weight compounds generally stimulate biodegradation of xenobiotics.

The role of oxygen in the degradation of xenobiotics merits special discussion. General participation of oxidative enzymes in the conversion of foreign compounds demonstrates the necessity for accessible oxygen in order to effect degradation of pesticide residues. However, studies of the degradation of applied pesticides in water and the soil of flooded rice fields have shed new light on the role of anaerobic processes in microbial degradation of xenobiotics.

Many persistent pesticides, reputed to be recalcitrant to microbial attack and thereby residing in soil for many years, are degraded much faster in tropical soils from flooded rice fields. Lindane was subject to anaerobic microbial degradation in muds and lake sediments. In addition, heptachlor, endrin, methoxychlor and dicofol also underwent intensive degradation in flooded soils.

The organophosphorus pesticide diazinon was eliminated from the flooded soil in two months, whereas its residence time in a non-flooded soil was six months. It should be noted that anaerobic conditions favour the accumulation of metabolites whose further decomposition requires the presence of oxygen. This phenomenon was observed during degradation of parathion, diazinon, carbofuran and other pesticides in soil.Anaerobic processes are highly dependent on the redox potential. Microbial degradation of xenobiotics is also related to acidity of the medium. This affects the rates of sorption of pesticides and their metabolites and also influences the processes of decomposition or the formation of conjugants.

ENHANCED DEGRADATION OF PESTICIDES

Repeated applications of the same biodegradable pesticides to soils can lead to a build-up of microbial populations capable of more rapid degradation of the applied pesticides thereby decreasing their efficacy in soil This is the case with many carbamates, acetamides, acylanilides and organophosphorus insecticides, herbicides and nematocides which upon application to soil undergo rapid degradation. The soils in which such pesticides produce no control of the target organisms are called the 'tired', 'problem' or 'aggressive' soils.

It was clearly demonstrated that frequent applications of 2-triazines, carbofuran, chloroethiocarb or other toxicants led to the complete and rapid loss of their activity due to their drastically enhanced degradation by the soil microflora. The isolation and identification of microorganisms from such soils identified the presence of strains capable of rapid metabolism of pesticides and their structural analogues.

This fact essentially restricts the efficient use of some pesticides in agriculture. Many factors appear to contribute to the emergence of problem soils. The most common are the concentration of organic matter in soil, pH, moisture level, temperature, aeration conditions, soil adsorption capacity, application rate and quantity of applied pesticides. To date, no systematic studies have been conducted to establish the number of pesticide applications after which the phenomenon of accelerated degradation of specific pesticides becomes operative in soils.

Laboratory studies of pesticide degradation kinetics in problem soils with parathion, pointed to a distinct enhancement of the degradation rate after the second or third application of the toxicant to a flooded soil. In the field, the degradation of carbofuran can show a drastic enhancement even after one application to sandy soils.

In practice, certain pesticides, for example herbicides, are usually applied only once a season, whereas certain insecticides are applied more frequently (2-4 times a season in rice fields and 10 or more applications tocotton). For various pesticide preparations, the threshold levels, that induce enhanced degradation and consequently activate the adaptive microflora, have different values.

Repeated applications of the same pesticide may, in some cases, slow down the rate of its degradation, e.g. alachlor and terbutryn. Certain treatments such as soil fumigation or solarization or treatment with various fungicides and bactericides have also been shown to affect microbial degradations and to slow down the degradation (ref. 31).

In the laboratory, experiments are generally conducted with higher concentrations of pesticides than those used in the field. This fact should be borne in mind when making recommendations for averting the development of problem soils and attempting to extrapolate laboratory results to field conditions.

The following techniques have been proposed as preventive measures against the development of problem soils: crop rotation, alternating pesticides, application of microbial inhibitors, new formulations which inhibit the degradation of pesticides, etc.. Of course, any of the proposed techniques will have some advantages and disadvantages. The ideal case will be to achieve a more effective control of pesticide residues in real environmental conditions in order to minimize the level of pesticide use and thus bring down the probability of development of 'problem' soils.

MICROORGANISMS ROLE IN FORMATION OF PESTICIDES

When a pesticide is in contact with soil for some time, a proportion of the compound or of its conversion products can become bound to the soil organic matter and cannot be extracted by normal analytical methods. The IUPAC Commission on Agrochemicals has defined soil-bound residues as: "chemical species originating from pesticides, used according to good agricultural practice that are unextracted by methods which do not significantly change the chemical nature of these residues.

These non-extractable residues are considered to exclude fragments recycled through metabolic pathways leading to natural products". The binding of pesticides to soil is a matter of some ecological and toxicological importance. Once bound to soil, it appears that pesticides are detoxified and lose their activity. However, it is not known if detoxification is permanent or temporary. Some studies indicate that certain bound residues can be released and subsequently perhaps taken up by plants or leached into the groundwater. The limited information available on the role of microorganisms in the formation and degradation of soil-bound residues of pesticides is critically reviewed below.

Formation of Soil-Bound Residues

Pesticide residues are often bound to soil organic matter. The rate of incorporation of pesticides into soil organic matter depends upon the type of pesticide and its subsequent transformation by microorganisms and abiotic factors. The extent of binding varies from 50 per cent within a few weeks to only a few per cent after one year. It is difficult to predict from the structure of a pesticide its potential to form non-extractable residues.

Microorganisms are involved in several ways in the formation of soil-bound residues of pesticides. For instance, they may degrade pesticides into conversion products that are reactive towards humus. A large number of pesticides such as phenylureas, phenylcarbamates, anilides, dinitroherbicides and certain fungicides are converted into halogenated and alkylsubstituted anilines which are bound to organic matter. Little is known about the mechanisms of formation of soil-bound residues of pesticides.

Many soil microorganisms have polyphenol oxidases which catalyse oxidative couplings. These organisms have been shown to form oligomers from typical humus monomers. Bollag and Liu studied the enzymic polymerization of humic monomers with 2,4-dichlorophenolr a breakdown product of 2,4-dichlorophenoxyacetic acid. They found that an extra-cellular laccase from the fungus Rhizoctonia practicola coupled dichlorophenol with several phenolic compounds, all of which were precursors of humus.

Oligomeric mixtures (up to pentamers) of 2,4-dichlorophenol and the humic monomers were also obtained. The same fungal enzyme also coupled pentachlorophenol and syringic acid to form several oligomers. 2- and 4-chlorophenol, 4-chloro-2-methylphenol, and 4-bromo- 2-chlorophenol, conversion products of 2- and 4-chlorophenoxyacetic acid and 4-bromo-2-chlorophenoxyacetic acid respectively were polymerized by the enzyme. Substituted anilines but not nitrophenols were also coupled.

Mathur and Morley observed incorporation of methoxychlor into a humic acid-like polymer by Aspergillus versicolor. Pure culture studies with Hendersonula toruloidea have indicated that substantial percentages of the ring carbons from 2,4-D are linked into humus-like polymers. The linked aterial is more resistant to decomposition in soil than the unbound compound.

There is evidence that for certain compounds like phenols, microbial activity of soil enzymes plays an important role in the formation of pesticide-bound residues.

This conclusion is supported by the fact that soil conditions which are optimal for microbial activity, such as addition of manure and higher temperature increase, the incorporation of pesticides in the soil matrix. Soil microorganisms may degrade pesticides into small fragments which are transformed into natural products.

Such fragments are, by definition, not regarded as pesticide-bound residues. However, it may be very difficult to distinguish between these natural breakdown products and relevant bound pesticide residues when C is found in exhaustively extracted soil incubated with a C-labelled pesticide.

Release of Soil-Bound Residues by Microorganisms

Once incorporated into soil organic matter, bound residues are generally rather stable. When 3,4-dichloroaniline was applied outdoors to soil in a plant-soil system, 46 per cent was still present, in the form of bound residue, two years after treatment. The turnover time for soil organic matter is quite slow and release of bound residues is probably gradual. However, several authors have obtained evidence that bound pesticides or theirmetabolites are released faster by the action of some species of microorganisms.

For instance, Hsu and Bartha found that small amounts of soil-bound residues of 3,4-dichloroaniline were mineralized to $C0_2$ upon incubation of soil containing bound residues with fresh soil. This mineralization decreased with sterilization or anaerobic incubation. Roberts and Standen reported that between 25 per cent and 40 per cent of soil-bound cypermethrin residues were mineralized to $C0_2$after 26 weeks of incubation of soil containing bound residues to which fresh soil had been added. Khan and Ivarson investigated microbial release of [^{14}C] prometryn from an organic soil. Extracted soil samples were incubated with a liquid soil inoculum. After 32 days, 27 per cent of [^{14}C] which has been initially unextractable could be extracted. No [^{14}C] was released when soil was incubated with sterilized distilled water.

Many of the workers observed the uptake by plants of soil-bound residues. It is possible that microorganisms which are active in the rhizosphere.e plant roots play a role in the release of bound residues from the soil. Also, increased microbial activity enhanced the release of $C0_2$from soil bound residues. This latter approach may be used to estimate the rate of mineralization and persistence of soil-bound residues.

Several examples of the release and breakdown of pesticide soil-bound residues by pure or mixed cultures of microorganisms have been reported. Mathur and Morley found that methoxychlor, incorporated into a humic material produced by the fungus Aspergillus versicolor, was subsequently released by the action of another soil fungus.

Khan and Ivarson incubated soil, containing bound residue of [^{14}C] prometryn, with different types of microbial populations. They found release of bound [^{14}C]-labelled compounds and $^{14}CO_2$. Hsu and Bartha studied the biodegradation of a humic-[^{14}C]-3,4-dichloroaniline complex isolated from soil, by the soil fungi Penicillium frequentans and Aspergillus versicolor.

The complex served as the only organic substrate for the fungi. Both fungi released $^{14}CO_2$ from the humic-dichloroaniline complex. As yet, it is not known whether studies with pure cultures may represent the actual situation in soils. It may be concluded that microorganisms are involved in the formation of soil-bound residues of pesticides either by incorporating intact compounds of metabolites into soil.organic matter, or by generating breakdown products which can react with soil components. Indirect evidence seems to indicate that under certain conditions, microbial activity can release and mineralize soil-bound residues. From the limited literature data available, no conclusion can be drawn about the potential environmental significance of the microbial release of bound residues of pesticides.

DEGRADATION OF XENOBIOTICS

Recent developments in genetic engineering of microorganisms have opened possibilities for the use of new microbial strains having enhanced degradative activity towards persistent pollutants. Genetic engineering techniques can be used to construct bacteria, capable of producing enzymes for pesticide detoxification.

At the present time, strains capable of degrading such persistent pollutants as mono- and di-chlorophenols, mono and di-chloro-benzoates, DDT and dicofol, 2,4,5-T, etc. have been constructed. These strains were constructed using strategies based both on engineering 'in vivo' (i.e. direct strain-to-strain transfer of an entire plasmid or of its fragments with subsequent selection of trans conjugant) and 'in vitro' techniques, i.e. gene cloning. The in vivo strategy proved to be successful for constructing the strain Pseudomonas aeruginosa BS827 with enhanced capability for degrading both DDT and dicofol. The same approach was used successfully in engineering bacterial strains capable of degrading mono- and dichlorobenzoates and 2,4,5-T. In fact, this strategy has opened up broad vistas for the construction of novel strains. However, the transconjugants produced are not merely strains with an increased number of plasmids but, as a rule, they contain complex genetic elements which lead to the establishment of new biochemical pathways that are difficult to predict.

The in vitro approach presents a number of advantages. It allows the selection of those genes which are indispensable for specific hybrid pathways while excluding the genes encoding unproductive steps of biodegradation. However, this method also suffers a substantial limitation. To successfully manipulate DNA enzymes, a complete knowledge of modern biochemistry and genetic organization of the initial metabolic pathways are required. The lack of such information on the degradative processes of most xenobiotics impedes greatly a wider use of this method.

One successful use of the strategy is the construction of a microorganism, based on the strain Pseudomonas sp. B13 which is capable of utilizing 3,4- and 3,5-dichlorobenzoates. Fragments of the TOL plasmid containing the genes xylD and xylL and DNA of NAH7 plasmid harbouring gene had also been cloned in the vector pBR322, which was subsequently ligated with the vector pKT231 for a wide range of Gram-negative bacterial hosts. The hybrid plasmids were introduced into the strain Pseudomonas sp.Bl3 to produce a dramatic increase in the metabolic potential of the strain enabling it to utilize 3,4- and 3,5-dichlorobenzoates and 3-, 4-, 5-chlorosalicylates.

Successful developments in constructing strains able to degrade persistent pollutants have brought to the forefront the problems of stability and resistance of such strains. Therefore, at present, one of the most urgent problems in this field, is the establishment of the factors which control the stability of a plasmid-carrying microbial population and the investigation of possibilities of effective use of constructed strains under natural conditions. Studies of the fate of genetically manipulated microorganisms in model ecosystems are especially important to predict their behaviour in natural ecosystems. The release of such microorganisms into soil or water systems, requiresspecial consideration in addition to their viability and genetic stability.

These embrace their impact on natural water and soil ecosystems, ability to exchange genetic material with the autochthonous microflora and the capacity for distribution over substantial distances and to occupy new ecological niches. Studies of the viability and stability of a number of strains which are interesting as potential subjects for subsequent genetic manipulations, revealed that individual strains were rather tolerant to abiotic stresses, remained viable under nutrient deficit conditions and could exist in the presence of antagonists. Constructed strains also proved to be fairly viable when released into soil.

The strain P.aeruginosa BS827 was active in the degradation of kelthane in soil and was persistent for several field seasons, whereas the plasmid introduced in the strain remained rather stable under selective conditions in the presence of the xenobiotic However, the lack of adequate genetic characteristics of the initial and constructed strains, scanty data on genetic interactions of microorganisms, especially on those engineered by genetic methods, under natural conditions, and imperfection of the monitoring of constructed strains released in the environment are becoming the cause of growing concern on the part of scientists and the public.

Some scientists are of the opinion that the use of such strains is not advisable in practice due to potential hazards to the biosphere. Genetically constructed strains appear to have some promise for the cleanup of industrial sewage, a step required for averting environmental pollution. Experiments on purification of industrial effluents from residual toxicants using the strain P.aeruginOs.a DC13 carrying the plasmid of biodegradation of a-methylstyrene, toluene and diphenyl showed that both the strain and the plasmid were stable and active during the entire 4-month experiment. In these experiments, it was also found that more than 99 per cent of the toxicant could be removed. Another direction where biotechnology may have a positive impact is the development of higher yields of microbial enzymes which degrade xenobiotics. To date, success has been modest. The enzyme parathion hydrolase has been shown to hydrolyse parathion:

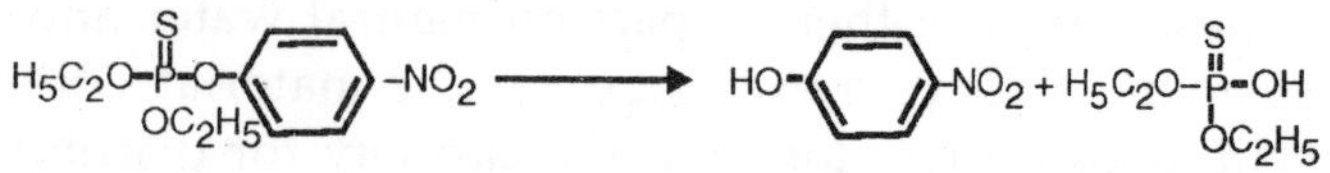

This enzyme, isolated from Ps.diminuta and Flavobacterium sp., displays a broad substrate specificity in relation to 0,O-diethylphosphorothioates. Both strains carry plasmids responsible for the biosynthesis of this esterase. In Ps.diminuta, the plasmid pCMSl has a size of 66 kb and in Flavobacterium 43 kb. The experiments on DNA hybridization showed a significant homology of the two genes responsible for the enzyme biosynthesis. Restriction analysis supported this homology. The work aimed at obtaining strains with enhanced capabilities for the enzyme biosynthesis is in progress.

PERSPECTIVES ON THE PRACTICAL UTILISATION OF MICROORGANISMS

Practical application of microorganisms for the degradation of xenobiotics is presently used almost exclusively for treatment of industrial sewage. Experience gained from laboratory assays and industrial uses indicates that both pure and mixed cultures of microorganisms are capable of degrading individual pesticides. This has been demonstrated for the degradation of 2,4-D, 2,4-dichlorophenol and parathion using pseudomonads. Immobilization of cells and enzymes opens interesting prospects for purification of industrial effluents from toxicants. Thus, an enzyme preparation capable of hydrolyzing nine organophosphorus pesticides was immobilized on fragmented glass which was packed in a column. The immobilized enzyme had a half-residence time of 280 days maintaining the specific hydrolase activity at the level of 0.035 to 0.5 pmolpr mg glass. The pseudomonad cells immobilized on glass wool were used to completely purify industrial effluents of such toxic volatile compounds asa-methylstyrene, toluene and styrene. The purification efficiency was over 99 per cent, and the cells displayed high activity during several months. Efforts to use active microbial strains for the degradation of pesticides and other xenobiotics under natural conditions have concentrated on two principal approaches.

The first one is the release of the active microorganisms into the soil and water systems which have been studied in model and natural ecosystems. In some instances, the desired results were obtained. However, the majority of experiments with genetically constructed strains were not successful, and the strains released into the ecosystems were either rapidly eliminated or did not exhibit any degradative characters. This approach calls for a more adequate investigation to eliminate or at least minimize the hazards associated with the uncontrolled distribution of genetically constructed strains.

The second approach, which also appears to be promising, attempts to activate the microflora of natural habitats by the introduction of appropriate inductive substrates, sources of nitrogen and phosphorus, and various cosubstrates. The US company "Sun Oil" has developed much experience in the cleanup of water systems, underground waters and limestones polluted with petroleum products by the introduction of dissolved nitrogen and phosphorus salts.

Finnish researchers have obtained positive effects in the purification of soils polluted with petroleum waste by the application of effluents from industries manufacturing yeast. Earlier, some work was mentioned that was aimed at enhancing the microbial degradation of lindane, DDT and other polychloroaromatic pesticides by ploughing alfalfa, straw and green pea mass into fields. The cleanup of pesticides in natural habitats by activation of their natural microflora calls for a more systematic effort.

Chapter 9

Laser Satellite System

LASER SATELLITE COMMUNICATIONS

As humanity moves into the third generation of telecommunications, it is also learning to deal with what many have come to know as the Knowledge Era, or the Information Age. As humans explore more deeply the knowledge economy, they are slowly realizing that it is business, not merely technology, vision, promise, or concepts that creates reality. At the centre of modern society lies not the dreamer, the politician, the scientist, or even the individual. At the centre of the economy, the community, the polity is the managed institution, including the modern business enterprise.

It is the modern business that embodies and exploits technology, information, productivity, innovation, and even the individual. The modern business, and the governmental, religious, social, and other managed institutions that support it, are modern man's way of accomplishing things. If laser satellite communication is to achieve its promise, it must become a business. These pioneer LSC companies have much more to do than simply building a business or executing a business model. Theirs is the challenge of creating an industry that empowers all of humanity to leap into the next generation of the telecommunication enterprise.

TECHNICAL SYSTEM CONCEPT

The nucleus of a laser satellite communication (LSC) system is a set of six spin-stabilized satellites in midlevel earth orbit (MEO) above the Van Allen radiation belts. A smaller set of three to five laser satellites in geosynchronous earth orbit (GEO) expands coverage, extends connectivity of the system to higher latitudes, and provides additional capacity for trunk, simplex, and non-real-time traffic. As they move around the earth in a circular orbit, the six intervisible MEO satellites would be deployed in two three-satellite laser rings offset from each other in longitude, while the GEO satellites would be positioned to achieve global connectivity and coverage of target markets.

Laser intersatellite links (ISL) would create a double laser ring at MEO linked to another GEO laser ring, both circling the earth. These interconnected laser rings would carry low-latency duplex and voice telecommunication traffic as modulated laser beams between the satellites. Laser uplinks and downlinks with regional ground stations would establish laser channels for transmitting traffic to, through, and from the laser rings and between the laser rings and any ground stations.

Through such a dual GEO-MEO constellation of laser satellites, modulated laser beams can carry digital and analog traffic between two terrestrial points at rates of gigabits—and even terabits—per second. The use of modulated—amplitude, phase, or frequency—laser reflection technology for LSC reduces the weight and size of satellites and substantially reduces the criticality of satellite attitude control. The resulting laser microsatellites are small enough that each telecommunication constellation of laser satellites could be deployed by a single launch vehicle.

Small satellites and boosters designed for responsiveness provide cost-effective capabilities for graceful system enhancement, replacement, or degradation. Binary failures of heavily invested first- and second-generation spacecraft and the ambiguous in-orbit failures arising from long-term effects of space weather degradation increase both the cost and the risk of orthodox telecommunication satellites. The long lead times of all heavy-payload launch schedules and the continuing arguments between launchon-need and launch-on-schedule complicate the simple replacement of an obsolete, degraded, or nonoperational satellite. For the third-generation laser satellite constellation, issues of replacement are simple and less stressful. Since the loss of a single satellite represents only a small fraction of the system's capability, replacement of a satellite will *never* be urgent! Because of the flexibility in choice of launch vehicles, schedules, and orbits for laser microsatellites, replenishment can occur with the efficiency of deliberacy, rather than the "cost-plus" urgency inherent in the crisis of subsystem failure in an orthodox RF satellite.

Many of the crippling disabilities of orthodox GEO satellites—orbital positioning conflicts, frequency interference, revisit delays, high costs and risks of spacecraft and launch, or latency—are largely negated by the powerful synergies of laser-linked MEO constellations.

The individual laser satellites comprise elements, subsystems, and technologies that are neither new nor different: cost and risk of subsystems and components are relatively low. Unlike the mega-LEO constellations, however—and in stark contrast to the local "footprint" of each first-generation FSS satellite—the laser satellite rings provide global connectivity with only a few simple microsatellites: complexity and sophistication remain in the ground segment. The novel, inverse architecture of a third-generation LSC system generates commercial might and technical promise in the responsive and pervasive presence of an entire system launched or stored on orbit, ready to communicate. "One satellite is cheap; many together are power."

This same technology, however, also increases the satellite's capacity to transmit information. The high frequency, capacity, and speed of laser beams using smaller, lighter antenna systems (telescopes) with lower power requirements make them an attractive alternative to the high-power RF signals operating at lower data rates in congested frequency bands with lower capacity.

The laser medium is optimally suited for use of wavelength multiplexing techniques to exploit the wide range of laser carriers for transferring multiple high-density messages simultaneously and efficiently.

PRINCIPLES

Any LSC system embodies the principle of modulating laser beams to carry information. The key to efficient communication over laser—rather than RF—lies in the extremely short wavelength of light—about a thousandth of that of RF. This difference allows the focusing antenna to be about a thousandth of the size and weight of the RF antenna and the data rate to be 1,000 times greater than the rate of RF transmissions. An optically communicating satellite can either be the size of a comparable RF satellite with 1,000 times the available capacity, or provide similar capacity in about a hundredth of the size of a comparable RF satellite. As with the well-known concepts of FSS communications, the LSC concept involves three basic elements:

- The space segment—design, features, control, and capabilities of the satellites; signal processing on board the satellite; means of launching them into orbit; the dynamics of the satellites' orbit; and the mechanics of positioning and attitude control;
- The signal segment—the carrier medium used for communicating to, between, and from satellites—uplink, intersatellite link, downlink; the effects of distance; sources of signal interference, attenuation, and instability; modulation schemes and protocols to ensure proper transmission and reception; effects of digitization and compression;
- The ground segment—placement, construction, functions, and operation of ground stations; antennae, receivers, transmitters, and other ancillary equipment; antenna engineering; multiplexing and multiple access schemes; tracking and telemetry control facilities; interface with terrestrial telecommunication networks and systems; administrative structure; managerial coordination of operations, finances, and business activities.

THE STATE OF THE ART

Since the mid- 1970s, both industry and governments, as

well as scientists, have explored the possibilities of laser-carried communications between ground stations, aircraft, and satellites. One of the most promising U.S. experiments is a demonstration LSC system developed by AstroTerra Corp. for testing on the second Space Technology Research Vehicle (STRV-2) launched by the U.S. Ballistic Missile Development Organization in 1998.

The AstroTerra system transmits 1.2 Gbps over a distance of 1,800 km, occupies less than 1,000 cubic inches, and weighs 14.3 kg. The satellite package comprises the transceiver telescopes and gimbals, an electronics module, and a deployment mechanism. The system employs single 5.4" Schmidt-Cassegrain receiver telescopes at both the satellite and the ground stations. Each of the eight communications lasers operates at 125 mW at a wavelength of 810 nanometers (nm) with a divergence of 80 microradians. Four of the communication lasers form one 600-Mbps channel with right-hand circular polarization, and four form the second channel with left-hand circular polarization. Depolarizers separate the communication signal into the two carrier channels for conversion—by two avalanche photodiodes—into an electrical digital signal for decoding. Power switching the laser diodes modulates the transmitted data onto the laser mean.

Two acquisition and tracking lasers—100 mW each at 852 nm with 500/1,500 microradian divergence—establish and maintain beam integrity. Cesium atomic line filters in the receiving telescopes separate the communication signal from the tracking beacon.

The receiving telescopes focus the acquisition-and-tracking signal onto a camera sensor to track it and drive the gimbal system. The powers of modern laser diodes and the magnifications of commercial telescopes are well able to transmit and receive coherent laser beams at interplanetary distances. The subsystem with the greatest risk is, thus, the pointing system. Since the radial dispersion of the laser beam has a Gaussian distribution, to maximize the power received the receiving telescope must point within ± 30 per cent of the centre of the beam.

Although the STRV-2 has a generous pointing error of ± 40 microradians, high precision bearings, sophisticated resolvers, and high-accuracy micro-positioners—piezoelectric drivers, flexible telescope lenses, or dual-beam tracking techniques—can achieve pointing accuracies within ± 1 microradian, which allow transmission divergences of as much as 3 microradians, well within the specifications of commercial laser transmission systems.

Although locating ground stations accurately in positions to minimize atmospheric attenuation and dispersion significantly decreases the pointing risk, it is always possible simply to increase beam divergence—by reducing transmission magnification—and either increasing transmission power or increasing receiver magnification to compensate for actual experienced pointing error.

THE SPACE SEGMENT

The LSC concept is innovatively contrary to both the orthodoxy of GEO fixed satellite service (FSS) and the current fashion of direct broadcast service (DBS) through either GEO or mobile satellites in low earth orbit (LEO) or MEO. Satellite engineers and operators of both first and second generations have become intrigued with, and remain under the spell of, the commercial possibilities of the unbalanced complexity concept. Large, powerful, complex, automatic satellites with many sophisticated, redundant, and fail-safe devices can communicate effectively with simple ground receivers.

Rather than facing a small, intensely competitive market of sophisticated ground station proprietors, operators of such high-performance—and very expensive—satellites can sell thousands of small, inexpensive, terrestrial cellular transceivers to a mass consumer market—clients, customers, or subscribers.

The trend to complexity imbalance is clear: whereas TELSTAR I weighed 172 pounds and carried a tiny whip antenna that could communicate only with the 85-foot dish antenna at Goldstone, California, modern satellites weigh several tons and can carry thousands of conversations between handheld cellular telephones. First-generation satellites use large, expensive, powerful satellites to broadcast to a large number of simple, inexpensive ground receivers, which may be mobile or stationary. The new second-generation LEO-MEO operators use more, smaller, lighter satellites with intensive digital frequency reuse and signal processing to reduce further the cost and size of ground receivers.

Both approaches require large amounts of transmission capacity in lower RF bands. Both approaches stress heavy investment in the space segment—either a few large, powerful satellites or large numbers of smaller, more sophisticated satellites—coupled with relatively simple ground segments. Laser satellite communication inverts the complexity imbalance by combining a small number of sophisticated ground stations with a small constellation of lightweight, inexpensive satellites in MEO, supplemented by a few more laser-equipped satellites in GEO.

Third-generation modulated laser retroreflection technology enables spinstabilized satellites to form an efficient six-satellite constellation of two laser rings in MEO above the Van Allen radiation belts, with a second smaller set of three to five satellites in the GEO. In addition to necessary systems for satellite housekeeping, each satellite carries several lasers equipped with LSC retroreflector subsystems to establish and maintain laser links with ground stations, with other satellites and with spacecraft in one continuous laser ISL. Each laser employs solid-state, phased-array beam steering linked to the satellite's retroreflectors to acquire and track other lasers and to maintain integrity of the laser links.

Piezoelectric modulation at 50-100 MHz of flexible tracking-beam retroreflectors at both transmitters and receivers—or even only at the transmitter —determines atmospheric distortion and generates instantaneous compensating distortion of the main beam. As turbulence or aerosols distort the counterdistorted main beam, the result is an undistorted signal beam focused on the receiver. Modulated tracking beam retroreflectors are particularly promising for reducing attenuation in a phased-array beam-steering system.

Through up- and downlinks, intersatellite links, and laser links with spacecraft, the LSC satellites receive signals—modulated laser beams carrying message traffic—from earth, from other satellites, and from spacecraft.

The receiving LSC satellite forwards signals to another satellite, or transmits them down to earth for delivery to an addressee. In addition, the uplink to each satellite may be used for beaming additional power to satellites for station keeping, pointing, or operating onboard systems.

Coordinated ground-station modulation-demodulation and onboard signal processing determines how a particular message is forwarded through the intersatellite rings and shunted into the proper satellite's downlink with earth. To receive laser-modulated communication traffic from other satellites outside the LSC system, each satellite operates an additional laser retroreflector subsystem linked to a solid state tracking and acquisition system. Guidance and processing signals—or reprogramming instructions—are transmitted on up- and downlinks through the patented retroreflectors on a single beam with traffic being carried.

THE SIGNAL SEGMENT

The signal segment of any LSC system is also in contrast to conventional first- and second-generationapproachesto satellitetelecommunications. Designers of geostationary RF satellite communication systems have historically increased transmission capacity through more transmitters, transponders, and receivers; narrower RF spot beams; higher radiated power; larger apertures and antenna systems; frequency reuse; and frequency coordination to allow collocation of satellites in the GEO.

The essence of the LSC signal segment is the simple set of satellite rings that effectively creates laser wired space. Use of modulated laser beams instead of RF broadcast cones to carry signals effectively minimizes the hardware on each satellite, reduces radiated power and interference, increases transmission capacity, and operates through small antenna systems on earth. In addition to high-latitude connectivity, the

GEO satellites provide additional capacity and alternative transmission paths. The MEO dual rings offer the low latency and strong signals needed for voice traffic. Two-way RF communications require two receivers and two transmitters, just as orthodox optical communications require two lasers, or lights, pointed at each other.

The LSC modulated laser retroreflector technology uses a single laser coupled with a laser reflector for each laser link both to acquire and track satellites and to receive and transmit information—and also reduces the criticality of satellite attitude control. Dual-frequency laser beams for message and acquisition functions and independent retroreflector modulation based on attitude determination maximize system simplicity, reliability, and capacity, while minimizing weight, volume, and power requirements.

The configuration of two offset MEO laser rings with a complementary set of laser GEO satellites allows each satellite to operate either as a repeater or as a laser source. Thus, signals can pass in both directions in the rings, as determined by routing and processing performed in the ground stations. The configuration of nine to ten satellites optimizes the balance between capacity and cost and provides low latency where required, general interconnectivity with other systems, and near-global coverage. Orthodox RF practice places uplinks and downlinks on different frequencies—each of which consumes transmission capacity—and requires a third frequency for acquisition and tracking.

In contrast, the LSC retroreflector system combines these functions in a single beam and doubles capacity with a dual-frequency configuration.

Simultaneous initial illumination of satellite retroreflectors with multiple laser beams of different wavelengths and comparison of the datamodulated reflections both multiplies capacity and allows real-time compensation for any atmospheric distortion of laser beams.

Use of the optical—rather than the radio—range of frequencies allows transmission of huge amounts—terabits—of data. The optical wavelengths of laser beams provide capacity beyond that of the entire RF spectrum. Exploiting the capacity of optical wavelengths for telecommunication traffic requires only a technically complex and sophisticated—but eminently practicable, inexpensive, and feasible—process of modulation-demodulation and conversion-reconversion between RF and light.

Orthodox RF antenna theory and broadcast technology involve manipulation of antenna size and shape around effective isotropic radiated power (EIRP), aperture, and a large propagation cone, which do not require fine beam pointing.

Laser satellite communication, in contrast, uses a highly collimated propagation beam at low power and replaces the large satellite antenna with fine pointing by phased-array beam steering directed by retroreflectors.

The ground antenna system can be reduced to a simple 1-meter reflector antenna. To complementhumanity's current fascination with low-capacity, low-speed, low-frequency, high-power broadcast, laser satellite communication provides high capacity and high speed in a high-frequency, low-power, point-to-point communication channel. The LSC system invites migration of high-capacity data and simplex video applications beyond the RF spectrum and exploits the inherent high capacity and high speed of the optical spectrum. The LSC signal segment based on retroreflector laser technology is simple, fast, inexpensive, and reliable.

THE GROUND SEGMENT

For any communication satellite system, the ground segment comprises fixed ground stations, mobile receivers and transmitters, and the satellite control facilities. In addition to the several varieties of mobile receiver-transmitters —airborne, marine, vehicular, and portable—fixed RF ground stations are of four types: high-capacity, mid-capacity, master, and receive-only. High-capacity stations are large, fixed installations that carry trunk communications for multiple large users. Midcapacity stations connect satellites to private networks carrying video conferencing, electronic mail, data transmission, and duplex voice communications.

Master ground stations link very small aperture terminal (VSAT) or ultrasmall aperture terminal (USAT) networks and terminals —carrying digital voice, video, and data—to a satellite. Receive-only ground stations are the least expensive and receive voice-only, data-only, or televisiononly.

Like the space segment, the LSC ground segment presents a contrast to the orthodox complexity imbalance of geostationary satellite practice. A few sophisticated ground stations—the minimum is three; a set of five provides redundancy—perform most of the complex functions of operating a communication system through laser links. The interfaces between the satellite system and local service providers in the connected terrestrial telecommunication networks involve several active, intelligent functions that must be performed at each LSC ground station:

- Making physical connections with the terrestrial networks or systems—public or private; circuit-switched or packet-switched; twisted or coaxial copper wire, fibre-optic cable, microwave, RF broadcast, or modulated laser beam—that provide and receive signals; these connections might be any or all of three distinct types:
- For an arriving terrestrial or space laser link carrying a modulated laser beam, the input interface would simply shunt the modulated laser beam into the uplink, while the output interface would simply shunt the downlink into the terrestrial laser link;
- For a terrestrial fibre-optic network or cable connection, the input interface would remodulate the received signal to the uplink laser frequency and shunt it, while the output interface would remodulate the downlink laser signal to the fibre-optic frequency and shunt it;
- For a wireless RE connection, the input interface would remodulate the received RF signal to the uplink laser frequency and shunt it, while the output interface would remodulate the downlink laser signal to the RF broadcast frequency and shunt it;

- Transferring received signals either to laser carrier for transmission to the satellite, or to the appropriate terrestrial carrier;
- Processing signals (modulation and demodulation, multiplexing, amplifying);
- Acquiring and controlling satellites and detecting when a satellite laser beam is degraded below acceptable levels or has disappeared entirely;
- Managing traffic (routing, determining if traffic is in queue during a degradation of the laser beam; determining the nature of traffic for prioritizing and rerouting; storing interrupted or delayed signals, applying appropriate traffic reroute algorithms—store-and-forward, close-hop rerouting, best-network search and forward, cancel, or return to sender—shifting traffic because of congestion; informing senders of errors, delays, rebroadcast, and other carrier states).

The LSC ground stations are most analogous to a multifunctional—highcapacity, mid-capacity, or master—RF ground station, since the LSC system can perform all three functions simultaneously: trunking to multiple users, distribution to separate users, and connection to a discrete network. As for any network of multiple satellites, a LSC network operations centre on earth controls all satellites, couples satellites and ground stations or mobile receiver-transmitters, and monitors system integrity and stability.

All of the intelligent functions of the LSC system occur in the ground stations to ensure that the signal arrives and leaves with no degradation, loss, or corruption. Since some satellites are not in geostationary orbit, each ground station—like each satellite—uses retroreflectors to acquire and track each satellite as it comes into range.

Since traffic may not be received from the terrestrial network as laser, each ground station remodulates terrestrial signals to laser frequencies and satellite signals to appropriate frequencies for transfer to terrestrial systems. Each ground station also performs some satellite and network control through a modulated retroreflector link. Intuition suggests the benefits of adapting existing RF ground stations to LSC requirements, much as the common antenna towers, arrays, and fields are able to accommodate different types, frequencies, sizes, and interfaces of various RF antenna systems. Although it may be possible, and clearly cost-effective, in some cases to adapt existing RF ground stations to the LSC system requirements, or even simply to lease space, such factors as location, laws and regulations, interface requirements, and protocols, suggest that such cases will be few.

Location

Sensitivity of laser beams to atmospheric interference—refraction or reflection—is the primary influence on location of ground stations. Lack of cloud cover is the most important criterion. This fundamental condition mandates location of ground stations in areas of low ambient atmospheric aerosols (clouds, airborne pollution, fog, smog, and haze). The ideal stations would be at high altitudes where the weather is dry and clear, regularly spaced around the earth's circumference, but not necessarily on the equator. Some intuitively obvious sites are Australia, Peru, Denver or Phoenix, and southern Spain. Mountaintop locations near astronomical observatories are also ideal. Some of the stations could operate unattended, subject only to scheduled maintenance and emergency repairs. Reliable high-bandwidth telecommunication links to the rest of the world are vital. Fibre-optic cable and terrestrial laser are the optimal media.

Cloud Cover

Distribution of cloud cover over the earth's surface is neither uniform nor even symmetric. Nephanalysis and meteorology have discerned a stable pattern of cloudiness related to latitudinal zones that encircle the earth. In both the northern and southern hemispheres, clouds form preferentially at the mid and low latitudes along all meridians. Minima, with high probabilities of extended clear weather—0 per cent-10 per cent cloudiness—and good seeing conditions —10 per cent-

30 per cent cloudiness—occurat the high latitudes and in the subtropics. Within these latitudinal zones, cloudiness is affected by local and regional orological, topological, and atmospheric features. In the Northern Hemisphere, during all seasons and over oceans and continents, cloud cover is minimal in the Arctic and the subtropics.

Over the continents, cloudiness increases with decreasing latitude from a minimum at about 80° N to a maximum between 60° N and 50° N and then decreases to another minimum at 15° N-20° N. The southern hemisphere shows a similar latitudinal distribution of cloud cover: maxima in the low (0°-10° S) and middle (55° S-60° S) latitudes, and minima in the subtropics and Antarctic at all longitudes. Over a year, the average cloudiness of the southern hemisphere is notably lower than cloud cover in the north. Although variations occur cyclically over the seasons, the zones of maxima and minima have remained stable since meteorologists began collecting cloud data in the early twentieth century.

The standard deviation of cloudiness is slightly higher in the Northern Hemisphere than in the south. Whereas Northern Hemisphere cloud cover seems most stable near the equator, in the Southern Hemisphere "minimum variability of f [cloudiness] is observed in the middle latitudes (from 40 to 60° S)." Further, in contrast to a clear latitudinal shift of northern zonal minima with the seasons—because of the northward shift of the inter tropical convergence(ITC) zone every summer—the southern middle-latitude zonal minimum lies during the entire year between 55° S and 60° S.

Daily observations have been combined with broad zonal distribution patterns of cloudiness to develop broad frequencies of cloud cover around an entire global latitude zone. In the northern middle latitudes (35° N-65° N) clear skies appear around the world with a frequency of less than 10 per cent in all seasons.

In lower latitudes (15° N-30° N) the frequency of global clear weather is 20 per cent-25 per cent. In all zones of the Northern Hemisphere, global clear skies are the least frequent weather pattern, except in polar zones where clear weather has the highest frequency (30 per cent-45 per cent in summer and 25 per cent-30 per cent in winter). In the Southern Hemisphere, the likelihood of global clear skies in the middle latitudes (45°S-65°S, the monsoon belt) is negligible. In the tropics and subtropics (0°-40°S) global clear skies occur with a frequency of 10 per cent-15 per cent.

The frequency of "good seeing conditions" is greater than 50 per cent in winter in the tropics and throughout the entire year south of about 20°S-25°S—except Antarctica, where at 80°S-85° S the probability of clear skies is 60 per cent-80 per cent in all seasons, but only about 45 per cent-50 per cent at 75°S. Continuing work of the International Satellite Cloud Climatology Project (ISCCP) has refined the latitudinal patterns into a set of probabilities of cloudiness in particular locations for most of the world.

Within the broad pattern of latitudinal zones of cloud cover, the ISCCP models identify general locations with low probabilities of cloud cover. Determining definite probabilities for a specified level of cloud cover remains beyond the capabilities of current data, models, and analytical techniques. Although historical details of cloud cover are available for active observation sites, extrapolation of these data to other sites without empirical observations is at best an inexact art of *ceteris paribus*.

Orography

Although the stability of zonal patterns of cloud coverage are sufficiently robust to allow valid generalizations, the clear asymmetry and anisotropy of global patterns of cloud cover demonstrate the powerful effects of local orology, topology, and atmospheric conditions. Beyond the effects of clouds, however, the short wavelengths of laser beams are sensitive to orographic interference. Desert areas generate dust with even slight winds, and vegetative cover around the ground site reduces atmospheric dust. Turbulence—caused by microthermal atmospheric fluctuations—can cause scintillation, beam broadening, spatial incoherence, and phase distortion.

The profile of a suitable LSC ground site, thus, includes low cloud cover, predominance of sunshine, low particle scattering, low turbulence, and low ambient humidity. Sites without high probability of clear skies, however, are least suitable for reliable laser earthspace communications.

Thus, issues of particle scattering, turbulence, and security lose relevance where the criterion of low cloud cover is not met. Initial results of the atmospheric visibility model (AVM) programme conducted by the Jet Propulsion Laboratory were not surprising. Since the height of atmosphere to be traversed and its aerosol content decrease with elevation, higher altitudes reduce atmospheric attenuation, and the earth's atmosphere is less transmissive at shorter wavelengths.

Further, a shorter path through the atmosphere decreases scattering and power requirements. The accordingly suitable areas for the LSC ground stations are arid, at high elevations, slightly vegetated, probably in inhospitable, harsh climates, and isolated from other human activity. Unless some other facility—a nearby RF ground station, an astronomical observatory, a weather station, or a scientific experimental site—coincidentally exists at such a suitable location, the harshness and isolation of the area impose additional costs and risks to build, maintain, and operate the new LSC ground station. An additional criterion for selecting multiple sites is low correlation of weather patterns between them, to minimize the probability that more than one site is obscured by weather.

Studies conducted in the 1980s determined that with only three sites, the joint probability of clear visibility for at least one site is 94 per cent; with four ground sites the probability rises to 97 per cent, and two sites have probability of only 83 per cent.

Site Selection

These criteria mandate establishment of laser satellite ground stations at points on the earth that are very dry and experience as little cloud cover as possible. Meteorological observation data since World War II, satellite sensor data, combined with the detailed data of the ISCCP predict such regions at high altitudes with steady winds and a mountain barrier creating a windward cloud shadow, or in equatorial regions of thermal atmospheric rise.

In addition to Arctic and Antarctic regions, such points exist in the southwestern United States; the Lesser Antilles; the Chilean Altiplana Desert; southern Africa; the Sahara Desert and the Sahel, Spain, and Middle East; the Indian Dekkan desert, the Kirghiz Steppe, and the Siberian Uplands; Australia; and the South Pacific.

Within the United States, the region of more than 3,200 hours of annual sunshine includes western Texas, southern New Mexico, most of Arizona and Nevada, and southern California. In addition, several points in Hawaii-particularly Mauna Kea—show clear skies probabilities greater than 90 per cent. In the Caribbean Sea, a region of low cloud cover—8 per cent-15 per cent—covers the Lesser Antilles from Guadaloupe through Trinidad and Tobago. In South America, the high Atacama Desert in northern Chile from about 18°S to about 23°S is the centre of a high, dry region that extends with cloud cover increasing from virtually 0 per cent near Quillaga, Chile, to about 10 per cent near Lake Titicaca in Peru and in the southern Atacama in Chile.

In addition, several points in trans-Andean Argentina have mean levels of cloud cover below 10 per cent. In southern Africa and Madagascar, a region of near-0 per cent mean cloud cover extends north from the Cape of Good Hope to about 11° S. The driest, brightest region—cloud cover near 0 per cent—includes Zambia, central Zimbabwe, southern Angola, and Swaziland. Much of Namibia, Botswana, South Africa, Lesotho, Mozambique, and Madagascar experiences mean annual cloud cover of up to 8 per cent. In northeastern Africa, the centre of a near-0 per cent cloud cover region is at the border intersection between Egypt, Libya, Chad, and Sudan—about 25°E, 21°N. This region extends north to the Mediterranean Sea, west to the Algerian border—about 10° E—south to Lake Chad—about 16°N—and east to the Red Sea, with an extension into Somaliland.

Northwestern Africa—Algeria, Tunisia, Morocco, Western

Sahara, Mauritania, northern Mali, and Niger—experience slightly higher cloud cover.

The entire Arabian peninsula experiences mean cloud cover under 15 per cent. A region of near-0 per cent cloud cover centers on Riyadh—47°E, 25°N. At the western end of the Mediterranean Sea, much of southern Spain and Portugal south of about 40°N experiences mean annual cloud cover under 10 per cent. In southwest Asia, a broad region of low cloud cover—12 per cent-20 per cent—extends from Turkey through Iran, Afghanistan, and Pakistan, into the Indian Desert southwest of the Himalaya Mountains.

Although data are less reliable in central Asia, the Kirghiz Steppe between the Aral Sea and the Tien Shan Mountains on the border of China shows mean cloud cover generally below about 15 per cent-18 per cent.

The Siberian Uplands, northern Mongolia, and Chinese Inner Mongolia also show mean annual cloud cover levels of under about 15 per cent. All of Australia north of about 30° S experiences lower than about 15 per cent mean annual cloud cover, with an arc of near-0 per cent cloud cover extending from the Great Sandy Desert in Western Australia through the Sandy Desert in the Northern Territory to the Great Dividing Range in Queensland. In the southwestern Pacific Ocean, a triangular region of low cloud cover—8 per cent-15 per cent—extends from the Caroline Islands—150°E, 8°N—in the northwest to the Tuamotu Archipelago—140°W, 20°S—in the South to about 115°W, 10°S in the northeast—nearly to the Galapagos Islands.

Regulatory Requirements

Although national and international laws and regulations are quite specific on licensing procedures, siting, construction, standards, and operation of RF ground stations, neither the International Telecommunications Union (ITU) nor any other international organization has addressed the quite different regulatory issues involved with laser communication—either terrestrial or satellite. Some national regulatory agencies—but not those agencies regulating telecommunications—have begun to address propagation of laser beams as a hazard, as a form of atmospheric pollution, and as a weapon. No national legislatures or regulatory agencies have considered modulated laser beams as a form of communication.

Despite the comprehensive legislative and regulatory treatment of RF waves and propagation, the electromagnetic and physical differences between radio waves and laser beams suggest that existing radio regulatory procedures and processes can be, at best, a loose model for management of laser communications.

Until international law addresses the issues of laser communications, laser communication operators must presume that any regulations imposed by host governments on laser ground stations will be *ad hoc*, probably negotiated, controversial, and clearly different from those imposed on RF long-line ground stations.

The major safety concern regarding any LSC system is that of optical injury to anyone who might accidentally wander through the beam. The geometry of the system limits this threat to crew and passengers on aircraft and spacecraft flying through the beam. The U.S. regulations and standards have been established by the American National Standards Institute (ANSI) in ANSI 2136.1, American National Standard for Safe Use of Lasers, and many have been adopted by other countries. For the 810-nm and 850-nm lasers used in the AstroTerra system, the Maximum Permissible Exposure (MPE) is 320 x 10^{-6} W/cm^2. (Since 810 nm is beyond visible range, the critical aversion response time is not applicable.) Since even a low-power 1-watt system constitutes an ANSI Class 4 laser, operators must take appropriate safety precautions at the ground stations.

Interface

The LSC ground segment must accomplish technical integration and ensure compatibility with both public and private terrestrial telecommunication networks, although it can be expected that the signal received and returned to a network will be digital. Any interface between the LSC ground station and a terrestrial network or channel involves at least two considerations: software and protocols; and equipment.

Software and Protocols

The combination of transmission control protocol (TCP) and Internet protocol (IP) (TCP/IP) has become the *de facto* standard for all Internet traffic and for much non-Internet traffic. Although the target markets and applications for any LSC system are not limited to Internet traffic, the growth of Internet and multimedia traffic suggests that laser satellite communications must incorporate the software and protocols to accommodate TCP, IP, TCP/IP, and TCP over satellite (TCPSAT) standards. Although the features of a particular protocol are transparent to laser technologies, the LSC system collectively, as well as each station individually, must accommodatethe protocols used by the terrestrial networks to which they connect.

Hardware and Equipment

The LSC ground station must feed electrical signals received from a first-generation metal network to a laser exciter coil as a charging pulse to generate a laser burst for transmission. To reconvert a received laser beam to an electrical signal for delivery to a first-generation network, the ground station must direct the received beam through a set of photoelectric diodes and amplifiers to generate electrical signals based on changes in beam intensity. Signals received from a fibre-optic network may usually be shunted directly into the laser transmitter.

Although laser satellite communication may need to design specific devices for the particular situation of each ground station, these technologies are commercially available, reliable, and proven. In contrast to RF transmissions, laser beams have extremely low signal-tonoise ratio (SNR). Thus, beyond any controllable errors introduced by lightelectricity conversions, correction of SNR-induced errors are not expected to require the costly, sophisticated error-recovery technologies required for intensive frequency reuse in RF systems. Although laser attenuation is insignificant—even at GEO distance-atmospheric interference and beam diffusion may reduce received energy.

Both ground stations and satellites will require self-selecting beam detectors and amplifiers to compensate for energy loss during transmission, which will vary as atmospheric conditions change. Because any amplification, however, introduces additional noise, ground stations will require internal filtering systems to recover the normally low laser SNR.

As with light-electricity converters, these sorts of technologies—amplification and filtration—are commercially available, reliable, and inexpensive. For signal routing, the LSC ground stations must be able to shunt traffic from an incoherent satellite link to a robust link. Technologies and devices for accomplishing the former as a refinement of circuit-switched, or token-ring, routing are commercially available and reliable, although laser satellite communication may need to refine standard devices for specific situations.

An alternative approach, where alternative satellites and satellite laser links are available, is packet-switched internal routing. As with the circuit-switched approach, standard, commercial equipment, techniques, and devices are available for packet-switched internal routing.

External message routing and recovery from system failure depend specifically on the routing procedures of a connecting terrestrial network. To participate in external routing performed in the connecting terrestrial networks, the ground stations must continuously monitor the entire LSC network to determine whether a received signal can be delivered. In accordance with arrangements with connecting terrestrial networks, the LSC ground station may interface in either of two ways: signal rejection, or delayed transmission—storeand-forward.

Using the signal rejection approach, when a ground station identifies the LSC system to a transmitting terrestrial network as nonreceptive, the routing protocols of the transmitting terrestrial network route traffic through other channels, either to another LSC ground station or through traditional RF satellite and fibre-optic cable operators.

This type of routing is appropriate for real-time and high-priority messages. Terrestrial networks—or even individual users-using this approach to routing may establish priorities that the LSC ground station would respect. Using the delayed transmission approach, the LSC ground station holds a received signal until it can be transmitted into the LSC system.

This approach —batch mode—is used regularly for transmissions that can be made in batches (e-mail, messaging or paging traffic, data transfer, or trunk video). This approach to routing is most efficient for very large volumes of traffic that require many hours of transmission over RF or cable channels. The devices, technologies, and software for any of these standard routing techniques are commercially available, reliable, reasonably priced, and readily adaptable to laser communication systems.

TECHNICAL ADVANTAGES OVER RF SATELLITE SYSTEMS

The system establishes a dual-coverage global network with only three to five GEO satellites and six MEO satellites. Additional satellites increase flexibility, improve reliability through redundancy, and facilitate expansion of the system by establishment of additional ground stations. The ISL are technically simple, optically stable, and physically reliable. The laser beam operates unhindered by electromagnetic or other interference in the airless vacuum of space. Laser intersatellite links are nearly free from propagation losses. Passive first- and second-generation RF satellites rely on terrestrial networks to interconnect uplinks and downlinks and achieve global coverage through hops with corresponding latency.

Intersatellite links obviate all hops and provide global coverage without increased latency. The MEO time-delay for signal transmission (one-tenth of a second) is substantially less than that inherent in RF GEO satellite systems (one-fourth of a second), is not critical for simplex traffic, and accommodates IP modulation forms. The cost of launching and positioning a lightweight satellite in MEO at about 3,500 kilometers (km) above earth is substantially lower—up to 40 per cent-than the cost of putting a large RF satellite into the GEO (35,000 km, or 22,300 miles). With few mechanical components, low power requirements, and small optical and electronic devices, these satellites are both smaller and lighter than conventional high-power, high-capacity GEO satellites.

Early projections of LSC satellite weight are from 200 to 300 kilograms (kg). GEO RF satellites weigh up to 3,800 kg. This light weight allows deployment of the entire MEO constellation from a single launch vehicle. The narrow laser beam focused directly on a single receiver provides high security from unwanted interception, makes jamming expensive and difficult, and allows a laser communicator to control not only the message content, quality, and nature of the signal, but the dissemination of the information that the signal carries.

TECHNICAL DISADVANTAGES OF THE LASER SATELLITE SYSTEM

The MEO satellites are *not* stationary with respect to the ground. They must be acquired and tracked by the ground station through some electro-mechanicaloptical process. The MEO is close to the Van Allen radiation belts. MEO satellites may require some shielding and deteriorate more rapidly than do GEO satellites. Laser uplinks and downlinks are subject to atmospheric interference. Laser transmissions may even lose beam integrity as the beam passes through atmospheric aerosol obstructions (clouds and rain). The laser beam is subject to large propagation losses. Laser beams lose from 0.5 to 1.0 dB in transmission, in contrast to radio wave energy losses of about 0.4 dB between the transmitter and a receiver. A laser satellite system using LSC retroreflector technology cannot broadcast to multiple or mobile receive-only stations. The system is necessarily duplex.

DESIGN AND CONSTRUCTION OF THE SYSTEM

Most of any LSC system's technologies, components, and

devices are available in commercial off-the-shelf versions. Design of a prototype would require further research in the areas of solid state steering of laser beams, tracking moving laser sources, and acquiring exogenous laser beams. Construction of an operating space segment would require focused engineering and development of unique devices and components for satellite attitude control, platform stabilization, and laser-based satellite control techniques.

In the signal segment, construction of any sort of LSC system will require further research in atmospheric interference, attenuation, and instability of laser beams; development of modulation-demodulation schemes and adaptation of protocols involved in interface with existing telecommunication networks; and determination of the effects of digitization and compression on modulation, demodulation, reliability, or carrier integrity at laser frequencies.

An effective LSC ground segment will require further research and development in acquisition and tracking of MEO satellites, interfaces with terrestrial telecommunication systems, and identification of ideal locations of ground stations to optimize the balance of atmospheric conditions and access to terrestrial telecommunication networks.

A BUSINESS MODEL

Although every business deserves its own unique identity, it is possible to speculate about the appearance of any LSC business. From the comfortable perspective of the ideal situation, it is useful—and rewarding—to try to describe its business concepts and strategies, its markets, its structure, its finances, and even to hypothesize about its success.

BUSINESS CONCEPTS

Any business involved in laser satellite communications—to conduct research, operate ground stations, launch and operate satellites, or provide telecommunication capacity to some target market—6associations with:

- A high-speed, high-capacity converged terrestrial fibre-optic network,

- A FSS system,
- A teleport operator,
- Terrestrial laser or microwave links.

Since the LSC operator will own the intellectual property associated with the space segment—and possibly that of the signal and ground segments—an attractive business concept is to lease or license either.

The LSC operator might lease laser bandwidth, much as FSS operators lease RF transponders, to terrestrial users, to FSS or MSS RF operators, or to fibre-optic network operators. Alternatively, LSC might license intellectual property to teleport operators or RF satellite operators to convert existing RF "gateways" to laser gateways. Several complementary operational business concepts, which are not mutually exclusive, are also intuitively apparent for LSC operations:

- Globalizing a regional fibre-optic network with at least 2 Gbps (OC-48) capacity;
- Bringing telecommunication service to a developing country with a dry, clear climate;
- Contracts with governments to augment their strategic RF systems;
- Satellite news gathering;
- Carrying simplex traffic from GEO and other overloaded duplex services.

Instead of leasing bandwidth to users and operators, the LSC firm might allocate bandwidth to a set of subordinate or associated businesses that sell or lease bandwidth to functionally specialized wholesale customers or clients.

Providing Connectivity

Laser satellite communication systems are well suited for connecting other high-performance networks, especially fibre-optic networks that are physically separated, areas that do not have an extensive terrestrial communication infrastructure, and terrestrial laser systems.

Fibre-optic Networks

Communications are earth-point-to-earth-point with a

terrestrial network at each point. The LSC system obtains signals through a node or terminal of a high-speed, high-capacity, converged fibre-optic network —OC-48 operating at 2 Gbps—collocated with the LSC ground station. The system delivers signals to a node or terminal of another network also collocated with the LSC ground station.

The network operators obtain signals from original senders, deliver signals to ultimate receivers, and perform retail marketing and sales. A generic prototype is the connection of one converged network or public switched telephone network (PSTN) to another converged system or PSTN in another country, region, or continent. The concept requires cooperative associations with terrestrial fibre-optic network operators at LSC ground stations.

Less Developed Countries

Communications are earth-point-to-earth-point. One point is a node of a rural-remote telephony network in a less developed country (LDC), which is linked to local, national, or regional telecommunication infrastructures. The other point is a terrestrial node of the global telecommunication network. The LSC ground station in the LDC may connect to an existing terrestrial telephone network, a terrestrial laser or microwave network, a local PSTN, or a VSAT or USAT system. One ground station (probably the station in the United States established under another business concept) operates the LSC satellites.

The host government, in cooperation with some international or foreign development agencies, develops, funds, constructs, and operates each LSC ground station in each LDC as a development project. The concept requires appropriate associations with the host government, a funding development agency, and terrestrial network operators (fibre-optic, laser, or microwave).

This concept has no generic prototype. The LSC system obtains signals from the LDC under the terms and conditions of the development project and provides connectivity through LSC satellites to the rest of the world. Signals could include point-to-point voice, video, and data telephony, but not broadcast transmissions.

A model involving LSC-associated network operators involves installation of local terrestrial networks with the LSC ground station as a node interconnecting local and global networks. A model involving LSC-associated satellite operators involves installation of local VSAT-USAT networks with the LSC ground station as a hub. In either model, each LSC ground station is developed and funded as part of a development project with, through, and for the LDC host government.

Terrestrial Laser Operators

Communications are earth-point-to-earth-point with a terrestrial laser link at one, at least, of those points. The LSC system obtains signals through the terminal of a terrestrial laser link with either a terrestrial network or an ultimate user. The system delivers signals to a node or terminal of another network or to another terrestrial laser link with an ultimate user. The terrestrial laser operator either obtains and delivers signals, or arranges to relay signals through terrestrial networks as appropriate and performs retail marketing and sales. The generic prototype connects news coverage of a disaster or athletic event through a temporary or mobile terrestrial laser link with a regional LSC ground station to a broadcast network's central editing facility.

An alternative prototype for dense industrial and urban environments is a terrestrial laser communications link to relay signals between users and a regional LSC ground station at very high data rates, effectively bypassing the telephone connections—and lower data rates—through copper wire and fibre-optic cable.

For both prototypes, a shared —or partial—fibre-optic cable could provide emergency connectivity with the LSC ground station at a much reduced data rate. The concept requires cooperative association with at least one operator of terrestrial communication lasers and possibly with a global news network operator.

Providing Capacity

The inevitable migration of users and applications requires additional capacity and bandwidth at frequencies beyond the RF spectrum. Laser satellite communication is the obvious—and, at the present levels of science, the only-medium that can solve the problem of scarcity of bandwidth at the preferred radio frequencies. Laser satellite communications are eminently able to provide capacity to both users and operators.

RF Satellite Operators

Communications are initially earth-point-to-earthpoint with a fixed satellite service (FSS) or LSC ground station at each point. Replacement FSS satellites are expected eventually to carry laser intersatellite link (ISL) capabilities for space-point-to space-point communications between FSS and laser satellites. The LSC system obtains signals through laser-equipped satellites of FSS systems, or through interconnected FSS and LSC ground stations.

The system delivers signals—through either LSC satellites or laserequipped FSS satellites—to some FSS ground station or to a laser-equipped FSS satellite. FSS operators obtain signals from original senders, deliver signals to ultimate receivers, and perform retail marketing and sales. The LSC system could connect to some RF satellite operator's own system either at the RF ground station or aboard the GEO satellite. The ideal structural interface connects the FSS system to the LSC system through signal processing equipment on FSS satellites to shunt signals between onboard RF transponders and laser transmitting and receiving equipment on the same FSS satellite.

Establishing laser intersatellite links between RF satellites and LSC satellites would not relieve the RF spectrum pressure on the RF satellite. In contrast, several RF transponders aboard the LSC satellites would allow the RF ground station operators to shunt non-real-time, high-capacity traffic directly into the LSC system by refocusing their RF uplink beams.

At the RF ground station, simple switches and modulators

might shunt traffic from the RF uplink, convert RF signals to modulated laser beams, and feed them into a laser uplink to the LSC satellite.

A reverse system would return received laser signals to the client ground station through a laser downlink and reverse processing. An alternate structure would connect one FSS ground station to one LSC ground station through terrestrial laser, microwave, or fibre-optic channels, and concentrate signal processing at the LSC ground station. A generic prototype connects one FSS ground station to the LSC ground station in another country. A generic prototype for a future constellation of satellites with ISL between GEO and MEO satellites would connect any FSS and LSC ground station to any other FSS or LSC ground stations serving other countries. This concept requires cooperative associations with FSS operators, and possibly with terrestrial fibre-optic or laser network or circuit operators.

Teleport Operators

Communications are initially earth-point-to-earth-point between a teleport at one point and another teleport or the LSC ground station at the other point. The LSC system obtains signals through a teleport collocated with the LSC ground station or connected to a distant LSC ground station through terrestrial laser, microwave, or fibre-optic channels. The system delivers signals to either another teleport or directly to a terrestrial PSTN operator.

A generic prototype of the LSC-teleport relationship is connection of one teleport to another teleport in another country. The concept requires either cooperative associations with teleport operators at LSC ground stations or cooperative associations with FSS operators providing capacity to teleport operators, and may involve cooperative association with terrestrial network operators.

Specialized Operations

The unique features of LSC provide a diffused focus for specialized business operations to meet specific requirements for high-capacity, high-speed, secure, communication channels. Two intuitive such opportunities are the space business and microwave replacement.

Government and Institutional Space Business

Communications are space-point-to-earth-point, earth-point-to-space-point, or even space-point-tospace-point to supplement government civil or military strategic RF communication systems—U.S. Deep Space Network (DSN), Military Strategic Satellite Communications Network (MILSTAR), or Defence Satellite Communication System (DSCS). The LSC system obtains signals from space exploration missions through an upward-looking laser mounted on the LSC MEO—or FSS GEO—satellites, or through terrestrial or FSS interfaces. The system delivers through the LSC laser ring directly to terrestrial receivers with laser downlinks, or through FSS, terrestrial laser links, or fibre-optic cable, or stores signals and data at some LSC ground station for later retransmission and delivery. The generic prototype is laser connection between a spacecraft and one of the LSC satellites.

The LSC system relays data, sensor information, and imagery from the spacecraft through an appropriate LSC ground station and some terrestrial linkage to the relevant user. The system also relays telemetry signals from the user facility to the spacecraft. The concept requires a dedicated onboard laser with acquisition and tracking capabilities controlled from earth, as well as a laser on the spacecraft.

Adaptation of RF Microwave Infrastructure

Communications are earthpoint-to-earth-point with at least one point connected to a converted or adapted microwave network. The LSC system obtains signals from the connected microwave-laser terrestrial network and from the global telecommunication network.

The system delivers signals either to the global telecommunication network or to the laser-microwave network for relay to the ultimate user. The generic prototype is connection of the LSC regional ground station to existing microwave networks that have been converted or adapted to terrestrial laser linkages. The LSC ground station would extend that network through LSC satellites into the global telecommunication network. The concept requires cooperative associations with a terrestrial laser operator and possibly with a terrestrial microwave operator.

TARGET MARKETS

Laser satellite communication provides very-high-capacity, very-high-speed carrier services to several specialized niche markets within the larger market for satellite carriers of telecommunication traffic. Not limited to any single type of traffic—voice, video, or data—LSC is ideally suited to "a customer or market where the overwhelming need is bandwidth, and loss of communication is acceptable. But doesn't put the customer out of business—they simply have to shift to a less functional system until the primary one returns." The most salient services that these sorts of customers, users, and markets demand are not particular applications, but connectivity to both terrestrial networks and wireless systems in combination with high transmission capacity. Across several broad functional sectors, the physical features of laser beams exclude, however, many real-time or interactive applications.

The many visible, functional, consumer applications for telecommunication services—RF or laser, wireless or wired—are the primary markets of the clients, subscribers, users, or customers of any LSC operator. Rather than reflecting particular functional sectors of the vast, fashion-driven telecommunication customer—or retail—markets, the features of the broad LSC or wholesale markets include:

- Connectivity between the LSC ground station and the ultimate user through either terrestrial laser communication links, fibre-optic links—single cables or networks—or some other terrestrial interface, operated by a cooperative associate, strategic partner, client or customer of the LSC operator;
- Long point-to-point communication links; primarily

international carriers; large land masses and regions—China, Russia, India, Latin America, Africa—may also be included;

- A substantial amount of long-distance and international Internet, simplex video, and data traffic;
- Data traffic that can tolerate some interruption or delay of service;
- Areas where geography, finance, or politics may interfere with laying land lines —copper or fibre-optic—or providing RF satellite ground stations;
- A primary requirement for high-rate data transmission, with a secondary demand for transmission of voice and video signals;
- Various and disparate types of unrelated ultimate users.

The critical scarcity of suitable—or preferred—frequencies in the RF spectrum has created an implicit, growing demand for capacity at higher frequencies. Just as terrestrial fibre-optic operators are struggling to expand capacity, increase speed, and extend connectivity, wireless and satellite operators feel the need to shift some types of traffic out of the congested RF bands that are better suited to other—more profitable, urgent, or valuable—applications and users.

Thus, the target markets for LSC system operators include both some communicators (originators and receivers of high-volume traffic) and most operators (carriers of traffic). Most sectors of the telecommunication market are feeling the increasing pressures of frequency scarcity toward migration. Some—but not all—of the functional subsectors reflect these features of LSC target markets. Both criteria appear most saliently in the specialized sectors of trunk television, business services, RF satellite operators, satellite news gathering services, Internet service providers and carriers, government agencies and military services, scientific users, and rural or remote areas.

Trunk Television

A growing, but not highly visible, sector is that of trunk

television, which involves neither immediate broadcast nor real-time urgency. Primary users of this service distribute or exchange scheduled programming—broadcast and cable television—and news collection of live events—satellite news gathering (SNG)-for later insertion in a broadcast. An appropriate business associate for laser satellite communications in this market might be the operator of a fibre-optic network, a GEO satellite operator, or even the operator of a television network or private corporate network.

GLOBAL RF SATELLITE OPERATORS

The primary application for laser satellite communications is to provide high capacity to operators of satellites that use RF transponders to carry non-real-time traffic at congested RF bands—C- and Ku-band. The LSC operator provides a high-performance, inexpensive link to allow the RF satellite operator to reallocate its own capacity to higher-value traffic that requires lower RF bands. Any LSC penetration of this market encourages the migration of suitable applications and traffic to higher frequencies, and will relieve congestion of the preferred RF bands.

Satellite News Gathering (SNG)

A specialized application for laser satellite communications—in cooperation with an appropriate terrestrial carrier—is provision of high-capacity transmission services to global news gathering organizations between remote news sites and central editorial sites. Many athletic events, concerts, political or other incidents and disaster sites require continuing video and audio coverage. Many such events occur where it is difficult and expensive to lay cables but feasible to install temporary terrestrial laser links.

A cooperative associate—or customer—of a laser satellite operator might establish terrestrial laser communication links to relay signals—audio and video—from the site to a local editing station. Permanent fibre-optic or laser connections with a regional LSC ground station would allow subsequent forwarding for transmission to global editing stations or broadcast facilities. Appropriate business associates for any LSC operator in this market might be providers of terrestrial laser communication links or fibre-optic cable operators, as well as a global news gathering organization.

Internet

Modulated laser beams are well suited to provide high-speed, high-capacity carrier services for some specialized sectors of the Internet—especially the 190-230 high-level Internet service providers (ISP) currently using the existing telecommunication networks to carry Internet traffic. Specific Internet applications might include cache data, synchronization, and network database update, electronic mail, network management, research, and voice telephony, as well as others that are yet to emerge.

An appropriate cooperative associate for any laser satellite operator in this sector would be an ISP or Internet access provider that also operates an extensive converged, terrestrial fibre-optic network. Laser satellite communications would establish high-capacity global connectivity, which lower capacity RF satellites cannot provide at all and which transoceanic cables can provide only at very high investment cost. With the same business model used by RF satellite operators, the LSC operator might operate either as a wholesaler or as a retailer through a teleport or gateway.

Government, Military, and Scientific Applications

Government and military users across the world require high-capacity, high speed carrier services for routine traffic, high-volume transfers of data and imagery—analog and digital—periodic reports, and other non sensitive, non-real time traffic. A laser satellite communication system is an ideal provider of high capacity, high-speed laser carrier services to scientists exploring deep space —especially NASA—as an alternative and complement to the overloaded NASA Deep Space Network (DSN).

As the space-faring countries of the world send more and more probes, missions, and vehicles into space, DSN will be unable to communicate with them as frequently as has been common.

Congestion and increasing demand for DSN mission time has forced NASA into the second generation approach of the telecommunication industry: using radio technology to relieve congestion. Beyond the development of new autonomy technologies —to reduce the need to communicate with spacecraft—under NASA's New Millennium programme, the global space exploration community will require additional high-speed, high-capacity transmission and relay channels to meet the communication requirements of accelerating space exploration programs. Approaches to these market sectors might be either as a prime contractor to government and military agencies or in cooperative association with a major prime contractor.

It is likely that any approach will also involve a cooperative relationship with an operator of a terrestrial fibre-optic cable or laser link. A secondary approach to this market might involve government research projects related to laser communications.

Rural and Remote Areas

Many rural areas are currently served only by microwave telecommunication relay networks or various cellular telephony systems. These microwave networks operate over a substantial infrastructure based on line-of-sight terrestrial links. Microwave relay is being replaced by fibre-optic cable in some areas, but in many regions, terrain and high cost prevent installation of cable and reduce the effective capabilities of the traditional cellular telephony systems. Inexpensive conversion of line-of-sight microwave networks to laser networks extending from a single LSC ground station may be feasible in some situations. The speed with which terrestrial laser communication links can be established suits them well for emergency services in remote areas with LSC connections to global communication networks.

A combination of terrestrial and satellite laser communications is also apposite for the many remote operating sites for oil, gas, and mining, which are within line-of-sight of each other, of the shore, or of prominent terrain features. Where RF microwave communications are difficult—near the horizon, over long distances, because of scatter or Fresnel diffraction—a combination of terrestrial laser and laser satellite communication can provide a very high-speed link with global connectivity. In many areas of active petroleum exploration—the Gulf of Mexico, the U.S.

West Coast, Venezuela, and the Arabian Gulf—and in some undeveloped regions—Peru, North and South Africa, South Asia, western China and Mongolia, and interior Australia—combinations of laser and microwave links have experienced availabilities and reliabilities greater than 95 per cent. The most salient users in this sector include exploration companies, government agencies, disaster management firms, nongovernmental organizations, and medical facilities.

Additional Markets

Some additional markets and applications that may offer opportunities to LSC operators include multimedia services, data broadcasting, digital imaging and telephony, alternative or replacement for microwave relay, scientific visualization, and electronic newspapers. Beyond the primary markets, the potential markets and applications for high-speed, high-capacity communication at low cost are limited only by man's imagination.

MANAGEMENT CONCEPT

The third-generation LSC business concept is much more than orthodox diversification. The unique business concept of developing cooperative associations with major firms in several complementary industries suggests adoption, with some refinements, of what has been called the *parenting approach* to management of a diversified business. In addition to managing the functional operations of the firm, the parenting approach includes a primary role for the management team in discovering, developing, and exploiting new opportunities—such as those of third-generation migration, specialization, and global interconnectivity. In a dramatic reversal of the conventional inward focus of management on technology, the parenting approach calls for a predominantly outward focus of managers at all levels on relationships and opportunities.

CORPORATE STRUCTURE

As appropriate for the broad concept extending over several market sectors, technologies, and industries, any LSC operator must establish a central corporate headquarters with a staff section, an operating division, or a focused cooperative association for each of the major business concepts. In addition to the normal corporate functions—general management, administration, finance, legal affairs, contracting, marketing, and business development—corporate headquarters establishes and operates the LSC space segment through a network operations centre. .

Corporate headquarters conducts both research and development relevant to all aspects of space segment, including design, production, and launch of the satellites; the signal segment (signal conversion and processing, laser modulation and demodulation, software interface, and telecommunication technologies); and the ground segment (siting, hardware interface, and site construction). In conjunction with its corporate associates, corporate headquarters must pursue new business development and opportunities for laser satellite communications.

Cooperative Associations

The preferred structure of the cooperative associations with terrestrial fiberoptic network operators and FSS satellite operators is that of a joint venture between the LSC operating company and another system operator.

As a first priority LSC operators should establish cooperative associations with at least one North American converged fibre-optic network operator, at least one foreign fibre-optic network operator, at least one operator of GEO satellites over North America, and at least one operator of GEO satellites over Asia and Europe. These cooperative associations provide laser communications and wireless access servicesthrough high-speed, high-capacitypoint-to-point transmission for satellite traffic that can tolerate delays in transmission—non-real-time.

Corporate Business Divisions

As a second priority, the operating LSC divisions for government, teleports, and institutional business must generate and enter into contracts to provide LSC services and transmission capacity.

These contracts not only generate current revenues, but often establish credibility and advantages for subsequent LSC business. Such contracts may require additional arrangements with operators of terrestrial network operators or GEO satellites, which should be integrated into the cooperative associations that are the primary LSC business priorities.

Internal Staff Sections

Lower priority business areas—rural telephony and terrestrial laser—will require at least monitoring and development of potentially productive relationships with organizations and individuals that are active in those areas. Corporate headquarters can maintain an appropriate presence in these areas through internal staff sections.

These staff sections must develop relationships and information involving the several lower priority business opportunities. These relationships can be expected to lead to additional business divisions, joint ventures, or other cooperative associations suitable to local situations, as integrated technological concepts develop.

STRATEGY AND INNOVATION

The foundation of the LSC concept—technical and commercial—is innovation. The LSC managers must deliberately abandon the products, applications and services, the processes, markets, distribution channels, and business models of the first and second generations. Instead, they must develop the novel concepts of laser satellite communications, nurture them, and use them to lead humanity into the third generation. The LSC managers must seek systematic improvement and innovation—both technical and commercial—and deliberately exploit not only of commercial successes but of the unexpected changes that laser satellite communication will inevitably bring.

This flexibility for creating and exploiting change enables LSC systems to create possibilities —and even contracts—for future opportunities.

The third-generation LSC managers have "to build a different tomorrow on a proven today... [and] to create the different tomorrow that makes obsolete and, to a large extent, replaces even the most successful products of today." Beyond establishing strategy for resource allocation and market focus, the responsibilities of the LSC management team involve not only operations, but discovering, creating, and nurturing new businesses, managing information flows across disciplines and markets, directing knowledge creation, and exploiting flexibility.

Rather than simply allocating internal resources, LSC management actively leads a corporate-wide discovery process into the third generation.

A nuclear LSC hierarchy of staff functions and business divisions supporting a set of cooperative associations creates, assembles, and focuses the new thirdgeneration knowledge that second-generation markets and industries do not have. The LSC strategy reflects the concept of stimulating the market through three-pronged investment in capital equipment, a regional and global sales and marketing network, and a hierarchy of salaried professionals.

Initial investment in capital equipment—satellites and ground stations—and a nucleus of professionals creates a powerful LSC capacity to change the human telecommunication environment by moving into the third generation. Parallel investment—through and by cooperative associations with established fibre-optic network operators and satellite operators—initiates and controls parallel processes of complementary, systemic innovation that the market alone would be slow to stimulate.

Any LSC operators must establish an aggressive and expansive business presence and ultimately operate in several diverse markets to provide high-speed, high-capacity, point-to-point communication for traffic, users, and applications that can tolerate delays in transmission—non-real-time.

In contrast to orthodox telecommunication firms and technologies, laser satellite communication is not, however, specific to any particular application. The equipment, software, and processes used by the LSC operator for any specific application or function are similar—if not identical—to those used for many other applications.

Any LSC business must sell capacity, or bandwidth, rather than applications. The LSC customers are the providers of communication services in the forms of applications; it is these providers that need LSC capacity. Any LSC business concept must eschew any exclusive dedication to a particular, current, or fashionable retail market for an application. Laser satellite communication is a wholesale service sold to providers of retail applications.

The orthodox business strategies of many mobile satellite services (MSS) —and some fixed satellite services (FSS)—operators involve alignments with equipment manufacturers in countries where they plan to offer service. These manufacturers' networks of distribution and support help to obtain requisite governmental consents and delineate service distribution channels.

In some contrast, a laser satellite operator should, instead, develop cooperative associations with operators of third-generation fibre-optic networks and with operators of global telecommunication satellite systems as primary parallel strategies. At a later stage, the LSC operator must expand the business concept to encompass similar relationships with terrestrial microwave and laser communication network operators, as well as with some specialized users.

The LSC operator does not offer service or product directly to individual users or consumers, but rather to and through terrestrial and satellite operators that require additional capacity. Thus, the problems of local distribution, regulatory compliance, and pricing faced by retail MSS operators are not major elements of the LSC business concept.

OPERATIONS

The LSC corporate headquarters must be the focus for management of the space segment. The cooperative association with a satellite operator may include management of the GEO space segment by the satellite operator, especially in the case of laser-equipped GEO satellites.

Since the cooperative associations with network operators will focus on the functions of particular ground stations, decentralized management of the ground segment would probably be most effective for routing information to these cooperative associations, with monitoring, coordination and integration by LSC corporate staff. Responsibilities and authorities for network management and business support efforts should probably be allocated between LSC corporate staff and the appropriate functional cooperative associations.

Adopting the terminology of the international satellite organizations (ISO) and the International Telecommunications Union (ITU), the internal LSC management functions involve five main systems functions for:

- Network management, including monitoring and control of network links and interfaces, resource management, performance monitoring, billing management, and other administrative functions;
- Management of the space segment, monitoring and control of spacecraft, operating and refining the interfaces between the LSC system and its connected terrestrial networks;
- Management of the ground segment, including monitoring and control of the ground stations and related subsystems, appropriate internal data communication functions to exchange management information among ground stations, satellites, and managed network elements;

- Management of routing information; information identifying LSC subscribers is distributed among the ground stations; each ground station uses two registers: one contains routing data of subscribers in its territory, whereas the second contains the information necessary to validate and route calls to all subscribers outside its region;
- Business support; service provision requires a dedicated software structure commonly known as an operations or business support system; it manages such important functions as billing and settlement policies including determination of traffic revenue sharing among connecting network operators, regional operators, governments, and other service providers representing the one-stop-shopping customer interface—subscription billing, customer service, and seamless interfacing.

CAPITALIZATION

The second-generation approach to capitalization—traditionally adopted by institutional and venture investors—stresses risk reduction in a narrow business focus. By developing specific new technologies, manufacturing a prototype product or system, and confirming its competitive advantage in the target market or application, a separate initial project reduces risk before developing the commercial structures and relationships of the business. The decisions about deployment, type of system, system design, and business concepts focus on the results and success of this initial phase of technology development.

The diversified LSC business concept and cross-industry scope of the LSC technologies contrast strongly with the focus of standard capitalization schemes on a specific technology, application, or market.

Development of any LSC business, thus, seems inappropriate for traditional first- or second-generation approaches to capitalization. An innovative third-generation approach involves alternative schemes for capitalization, each built around a particular consortium of different types of investors, with different goals, from different industries, reliant on different technologies, and focused on different markets for different applications.

Each approach envisions the LSC operator as an autonomous operating business capitalized through four sources of capital—institutional investment, private investors, venture capital, and corporate investment—over several years.

A small initial investment by an institutional investor provides a robust structure for developing an operating consortium among investors from several industries. Small investments by respected and experienced telecommunication executives provide the aggressive, dynamic, innovative leadership that the LSC concept requires and deserves. The venture capital investor exploits the strengths and experience of the venture capital industry in developing emerging technologies.

The LSC Consortium

Each LSC consortium includes at least one major telephone company or terrestrial network operator, at least one satellite operator, a financial institution, and involves a large aerospace vendor.

Some also involve many other firms —equipment vendors, software houses, publishers, new networks, and even governments. In this third-generation extension of industrial and financial trends, any LSC entrepreneur should form such a group of cooperative associates from the industries that complement—or perhaps constitute—the global satellite telecommunication industry. The corporate satellite and network operators are critical operational members of any LSC consortium as both primary investors and strategic allies.

Corporate investors are primary prospects for cooperative associations; they also provide an important connection between LSC and the retail customer markets that ultimately generate the demand for capacity. In conjunction with technical development—perhaps under the sponsorship of a venture capital investor—the LSC entrepreneur develops through and with its corporate associates the commercial infrastructures, ground stations, and market presence to operate the business.

The combination of four sources of capital—institutional investment, private investment, venture capital, and corporate investment—serves well both to establish the LSC firm as an operating business and to introduce the thirdgeneration LSC concept to a second-generation world. An institutional investor —investment bank—would be most suitable to support the firm's initial efforts to develop a consortium of private, venture, and corporate investors.

At least one individual—private investor—with knowledge, experience, and expertise in the telecommunication satellite industry should be invited to invest a small amount and assume executive or technical responsibilities with the LSC corporate entity. At least one venture capitalist investor should be invited to support development of the technical concepts of the LSC system—space, ground, and signal segments—to test and demonstrate those concepts and to produce and deploy the LSC satellite constellations.

The essence of consortium financing lies in the distinct and appropriate cooperative association—joint venture, equity sharing, strategic alliance, or a specific, unique arrangement—with each selected corporate associate. In addition to major investment, the corporate consortium members provide supplemental technological capabilities, internal production capacities, specialized managerial skills, or specific market positions and presence.

Alternative Approaches

Five alternative approaches to business development and capitalization seem intuitively apparent: deployment of the complete GEO-MEO system; deployment of a small GEO system; deployment of a large MEO system reinforced with complementary laser devices installed on other operators' GEO satellites; deployment of a small unreinforced MEO system; and deployment of only a set of ground stations to operate through laser devices installed on other operators' satellites.

The Complete GEO-MEO System

The complete-system approach leads to two interconnected constellations of LSC satellites in GEO and MEO. The most expensive alternative, deployment of the complete system includes a GEO constellation of three to four satellites to support the business concept of providing transmission capacity to FSS operators seeking to migrate traffic to higher frequencies. The MEO constellation of six satellites provides low-latency connectivity between fibre-optic network operators seeking to extend connectivity beyond terrestrial connectivities. The complete system involves four to five LSC ground stations and at least two corporate associates, as well as an institutional investor, a venture capitalist, and several private investors. With adequate capitalization and cooperation, the LSC operator should be able to deploy the complete system within three years of initial capitalization.

The GEO System

Deploying only the small LSC GEO constellation with four or five ground stations is a lower-cost alternative. This approach assumes the ability to install the appropriate signal processing, interface, and laser communication devices on another operator's LEO, MEO, or GEO satellites to establish laser links with LSC GEO satellites and up- and downlinks with LSC ground stations. This approach requires LSC ground stations only where the other operators do not have them, but assumes an ability to shunt LSC signals through other MSS and FSS systems and to receive signals in space directly from other MSS and FSS systems.

Connectivity with other systems may be through intersatellite links, terrestrial connections, or up- and downlinks. Interconnected transmission capacity and speed would necessarily be limited to those of the associated satellite operators. The primary target market for this small GEO system is terrestrial converged network operators seeking global connectivity to other converged networks. The GEO system would involve four or five LSC ground stations, at least one converged network operator as a cooperative associate, an institutional investor, a venture capitalist, and several private investors. As for the complete GEO-MEO system, the LSC operator should be able to deploy the GEO system within three years of initial capitalization.

The Reinforced MEO System

This less expensive alternative assumes the ability to install the appropriate laser communication devices in some modular configuration on another operator's geosynchronous satellites to establish laser links between them and LSC ground stations. Each GEO laser module provides full interconnectivity between the other operator's FSS system and the LSC MEO system. Like the complete GEO-MEO system, a complete version of this approach would also require establishment of at least two ground stations within the range of each laser-equipped GEO satellite and laser intersatellite links between the LSC MEO satellites and the other operator's GEO satellites.

A minimal structure might omit intersatellite links and GEO-specific ground stations and operate the LSC MEO system and the laser system carried by the other operator's GEO satellites as separate systems connected only through the LSC ground stations. The Reinforced MEO system involves four or five LSC ground stations and at least two corporate associates—a GEO satellite operator and a converged network operator—as well as an institutional investor, a venture capitalist, and several private investors. The LSC operator should be able to deploy the Reinforced MEO system within three years of initial capitalization.

The Small MEO System

This inexpensive alternative deploys only the dual laser rings in MEO with the ground stations necessary to operate them. Like the complete GEO-MEO constellation, the small MEO system is not dependent on other satellite systems or operators. Providing low-latency global connectivity, the small MEO constellation of six satellites would focus operations on fiberoptic network operators seeking to extend connectivity beyond continental borders. A complete global system would involve four or five LSC ground stations and at least one corporate associate, as well as an institutional investor, a venture capitalist, and several private investors. With adequate capitalization and cooperation, the LSC operator should be able to deploy the small MEO system within three years of initial capitalization.

The Ground System

Deployment and operation of this initial small groundstation system would demonstrate the technical and commercial feasibility of laser satellite communications, as a robust low-risk foundation for developing and deploying a complete LSC system as a later project.

Elimination of the entire space segment obviates the costs of construction and launch of the satellites and reduces the capitalization requirement considerably. Completely dependent on other operators' satellite systems, this approach assumes an ability to install the appropriate signal processing, interface, and laser communication devices on another operator's LEO, MEO, or GEO satellites to establish laser up- and downlinks with LSC ground stations.

This approach requires LSC ground stations where they can maintain connectivity with other operators' laser-equipped satellites.

The ground-station approach assumes an ability to shunt LSC signals through other MSS and FSS systems and to receive signals directly from other MSS and FSS systems through terrestrial connections, or up- and downlinks. The systems' transmission capacity and speed would necessarily be limited to those of the associated satellite operators.

The ground-station system involves four or five LSC ground stations and at least two corporate associates, as well as an institutional investor, a venture capitalist, and several private investors. The LSC operator should be able to deploy the ground-station system within three years of initial capitalization.

Capitalization Stages

Each of the alternative approaches involves a different combination of investors and stages. For all approaches, the first stage relies on an institutional investor and a private investor; the narrow focus of corporate activity would be to arrange capitalization and business relationships for the second and third stages.

The second stage would introduce a venture capital investor to develop the scientific concepts and technologies to deploy the LSC system. The third stage would involve corporate associates with an emphasis on development, production, and deployment of the system, establishment of markets, development of interface infrastructures, and construction of ground stations.

INVESTMENTS AND REVENUES

Like every start-up company, any LSC business can expect no revenues for several years. After three to five years of development and production, the launch of the LSC satellites would herald the beginning of revenue flows.

Prelaunch Investment

The LSC consortium allocates most of the investment capital to development, production, and deployment of the LSC satellite system with four or five ground stations. After an initial period of organizing the company, the LSC managers would allocate investment capital to the design of the satellites. Subsequent second-year investments would be allocated, in addition to contracted development of satellites, to developing the signal segment and its interfaces with terrestrial networks and other satellite systems, and to the design and location of ground stations. The highest expenditures of investment capital can be anticipated in the third year for production and launch of satellites and construction of ground stations.

Index

S

T

U

V

W

X